ALSO BY BERNIE KEATING:

When America Does It Right.
   AIIE Press, Atlanta, GA, 1978

Riding the Fence Lines: Riding the Fences That Define the Margins of Religious Tolerance.
   BWD Publishing LLC, Toledo, OH, 2003

Buffalo Gap Frontier: Crazy Horse to NoWater to the Roundup.
   Pine Hills Press, Sioux Falls, SD, 2008

1960's Decade of Dissent: The Way We Were.
   Author House, Bloomington, IN, 2009

Songs and Recipes: For Macho Men Only.
   Author House, Bloomington, IN, 2010

Rational Market Economics: A Compass for the Beginning Investor.
   Author House Publishing, Bloomington, IN, 2010

Music: Then and Now
   Author House Publishing, Bloomington, IN 2011

A ROMP THRU SCIENCE:
Plato and Einstein to Steve Jobs

By Bernie Keating

authorHOUSE

AuthorHouse™
1663 Liberty Drive
Bloomington, IN 47403
www.authorhouse.com
Phone: 1-800-839-8640

© 2012 Bernie Keating. All rights reserved.

No part of this book may be reproduced, stored in a retrieval system, or transmitted by any means without the written permission of the author.

Published by AuthorHouse 7/12/2012

ISBN: 978-1-4772-2382-6 (sc)
ISBN: 978-1-4772-2380-2 (hc)
ISBN: 978-1-4772-2381-9 (e)

Library of Congress Control Number: 2012911416

Any people depicted in stock imagery provided by Thinkstock are models, and such images are being used for illustrative purposes only.
Certain stock imagery © Thinkstock.

This book is printed on acid-free paper.

Because of the dynamic nature of the Internet, any web addresses or links contained in this book may have changed since publication and may no longer be valid. The views expressed in this work are solely those of the author and do not necessarily reflect the views of the publisher, and the publisher hereby disclaims any responsibility for them.

Science?

Many things are already known:
    Sciences of consequence,
        And those that nurture our soul.

How did they do it,
Discover those things?
    Isaac watching his apple,
    Albert at his kitchen table,
    James Watson tinkering with the basis of life,
    Steve Jobs pushing the envelope.

Are we only particles,
A nuclear field,
A philosophical illusion?

No matter; we are here,
    Still questioning,
        And relishing life.   Bernie Keating

TABLE OF CONTENTS

Preface ix

1	The Night Sky: Cosmology	1
2	Evolution: Darwin & Leakey	7
3	Psychology and You	13
4	The Science of Motion	21
5	Mathematics	27
6	Heredity and the Genome	31
7	Chemistry: Marie Curie & Linus Pauling	37
8	Electronics	43
9	Computers	47
10	Physiology, the Human Body	59
11	Nutrition	73
12	Physical Fitness	81
13	Economics	87
14	Weather & Climate	93
15	The Science Involving Energy	99
16	Water Science	111
17	Space Exploration	117
18	Sociology	121

19	Physics: The Next Five Chapters	127
20	Einstein and Gravity	129
21	Electromagnetism	137
22	The Nuclear Bomb	143
23	Nuclear Weak Force	147
24	Scientific Research	149
25	Philosophy	155
26	Summing Up	161

End Notes 163

PREFACE

Science is a marvel of modern life. We know how to fight disease, brighten-up our homes, communicate instantly around the world, and a host of things that have improved our quality of life. Most have been discovered during the past millennium. How did man accomplish this and who blazed the trail?

When I graduated from college sixty years ago with a degree in physics, I could claim (tongue-in-cheek) the title of a "scientist". Although I practiced only in the margins of the profession as a manager, my research in college qualified me for membership in the Sigma Xi Scientific Honorary, and I have diligently studied all their journals during the many years since and kept knowledgeable about scientific research.

I am fascinated by discovery: who discovered what, and how. My interest ranges from a look outward at the night sky with scientists such as Kepler, to physicist such as Einstein, an inward look at psychologists such as Maslow, and philosophers such as Plato.

Since I've forgotten considerable during the past six decades, I took myself back to school with research to rediscover some of the answers. Please join me.

Bernie Keating

THE NIGHT SKY: COSMOLOGY

As a boy in Buffalo Gap during the hot summer nights, I often slept outside under the stars and wondered what there was in space. Five centuries before my time, the young boy, Johannes Kepler, did the same in his far-away Germany, and a century earlier than that in what is today Poland, Nicolaus Copernicus also star-gazed. As early as twenty centuries ago, the little Ptolemy boy in Ancient Greece did the same. All of us have been fascinated with the question: what is out there and where do we fit in?

As each new scientific theory came into existence, each scientist stood on the shoulders of those who came before in an attempt to put a framework on what they saw as they looked into the night sky. Each did it within the cultural environment in which they lived; their social and religious environment had a profound impact on how they perceived the outside world.

Claudius Ptolemy, who was an ethnic Greek and a Roman Citizen living in Alexandra, Egypt, looked into the night sky in wonderment at the stars. The result became his theory that the universe was a set of "nested spheres" with the earth at the center. Few details of Ptolemy's life are known for certain. His writings were in Ancient Greece and he is known to have utilized earlier Babylonian astronomical data. Ptolemy's model of the universe was the authoritative text on astronomy for over a thousand years through the Middle Ages. It was eventually replaced during the Renaissance by the Copernicus theory. [1]

Nicolaus Copernicus was born in 1474 AD in Prussia, part of the Kingdom of Poland. The theory of the universe that he developed was a considerable advance over the theory of Ptolemy and came to be known as the "heliocentric" theory. With few exceptions, some of which sprang from the religious climate in which he lived, his theory of the universe is essentially the same as we have today. He had the sun as the center of the universe (which we now know is true only for our planets), but he correctly stated that the earth and planets revolved around the sun, that the earth performed a complete rotation on its fixed poles, and explained that the daily movement of planets and the sun in our sky was the result of the earth's motion

It is clear that in the 16th century Copernicus had things figured-out almost entirely correct. The exceptions were a few nuisances (such as "the highest heavens abide") that he included to be in concert with his Roman Catholic faith since it was important that his theory could be successfully explained to Pope Clement. After hearing the theory explained to him by a couple Cardinals, the Pope was pleased and gave it his blessing.

How was Copernicus able to develop his theory with the information then available? It is truly amazing, and that is why he has gone down in history as one of the most important astronomers of all time. A century later, Johannes Kepler was to stand on the shoulders of Copernicus as he took theory a step further in explaining the trajectories of planets through space.

I have always been fascinated with Kepler and how he went about developing his theory that predicts with great accuracy the travel of planets and even the travel of spacecraft in modern times. How was he able to develop those formulas?

Little Johannes Kepler lived in the German state of Baden-Wurttemberg. My own personal connection to Kepler is that my paternal great-great grandfather came from this same region, and since the first two letters in our last name are the same, perhaps there are some of those same genes in my make-up. Kepler's father was somewhat unstable and earned a precarious living as a mercenary and left the family when Johannes was five years old. (oops, I hope the connection between our family genes stops there.) Becoming a single parent, Kepler's mother, somewhat of a free-spirit, was a healer and herbalist who was later tried for witchcraft. [2]

Kepler arrived on the world scene during the era of the Renaissance -- a time of great change and "rebirth". The Protestant Reformation was already a major challenger to the Roman Catholic Church and religious argument was a factor in all aspects of social and scientific life. The Thirty Years War between Protestants and the Catholic Church occurred during Kepler's lifetime. At that time he was living in Prague and was an advisor to Emperor Rudolph of the Austrian Empire where the only acceptable religious doctrine was Roman Catholic, but Kepler's position in the imperial court allowed him to practice his Lutheran faith unhindered. [3]

While it had little impact in the interior of Europe where Kepler lived, this was also the era of discovery and exploration of foreign lands. Sir Walter Raleigh was in his world-wide travels and the Jamestown colony was established in 1607 in Virginia in the "new world."

Kepler's education in the eclectic fields of theology, philosophy, mathematics, and astrology would serve him well in a study of the universe. Early in his career he had the desire to become a minister. [4]

His interest in planetary motions came with an epiphany [5] when a periodic conjunction occurred of Saturn and Jupiter in the zodiac and he realized that their paths included one inscribed and one circumscribed circle of definite ratios. This could be no accident. As a mathematician, he thought there might be a geometrical basis for the universe. He began experimenting with 3-dimensional models of the six known planets -- Mercury, Venus, Earth, Mars, Jupiter, and Saturn. However, despite many successes in scientific discoveries in other areas, he was unable to carry this work on his planetary paths to completion for another twenty years from 1595 until 1615 A.D.

During the intervening years he did considerable other work: published manuscripts focusing on optical theory; described the inverse-square law governing the intensity of light; reflection by flat and curved mirrors; principles of pinhole cameras; astronomical implications of optics such as parallax; and the apparent sizes of heavenly bodies. He also extended his study of optics to the human eye and is credited as the first to recognize that objects are projected inverted and reversed by the eye's lens onto the retina. [6]

In 1615 at the age of forty-six, Kepler published his Laws of Planetary Motion. Here are statements of these three laws: [7]

> One: *The orbit of every planet is an ellipse with the Sun at one of the two foci.*

Two: *A line joining a planet and the Sun sweeps out equal areas during equal intervals of time.*

Three: *The square of the orbital period of a planet is directly proportional to the cube of the semi-major axis of its orbit.*

Kepler was a superb mathematician and was able to support each of his laws with the mathematics needed to define them.

So, what are the scientific advances that came with the laws? Perhaps the most significant are those of Galileo and Newton who went on to formulate the "Laws of Nature", which govern the behavior of all matter and energy. An example of a fundamental principle that was proposed by Galileo in 1642 and extended by Einstein in 1905 is the following: "*All observers traveling at constant velocity relative to one another should witness identical Laws of Nature.*" From this principle, Einstein derived his Theory of Relativity. [8]

Fifty years after Galileo, Isaac Newton described Universal Gravitation and Laws of Motion. He established that Kepler's laws of planetary motion were consistent with his theory of gravitation. Notwithstanding the story about the apple falling from the tree that hit Newton on the head (which became his epiphany for discovering gravity), it is clear Newton utilized Kepler's laws to advance his own theories.

Since the time of Copernicus and Kepler there have been many other advances in our knowledge about the universe. For the most part these came with the development of more powerful detection instruments such as the 100 inch telescope on Mount Wilson in California.

As a personal aside: in 1941 on a family trip from South Dakota to California, a highlight of our trip came with a visit to see this telescope on Mount Wilson, followed by a trip to Pasadena to view the 200 inch telescope lens then being polished in a lab at California Technical Institute, later to be installed on Mount Palomar. Thirty years later when I was a scoutmaster in Palos Verdes, my scout troop climbed from San Marino up the side of Mount Wilson where we had an over-night campout. The trailhead was a few blocks from the Rose Bowl, and with a good set of binoculars from our campsite you could watch a football game -- the ultimate locale for "Tightwad Hill". (Alas, since this was summertime, there was no football game in the stadium, so nothing to look at other than the chaparral vegetation.)

In 1929 the prevailing view of the cosmos was that the universe consisted entirely of the Milky Way Galaxy. Edwin Hubble, a thirty-five year old scientist, fundamentally changed the scientific view of the universe by proving there were other nebulae outside our own Milky Way. Then Hubble made an even more dramatic finding. He formulated the Red Shift Distance Law of Galaxies, nowadays termed simply *Hubble's Law*, which established the theory of an expanding universe (hence -- the Big Bang theory). For this, Hubble received a Nobel Prize. [9]

Then in 1998 cosmology was shaken to its foundations again as two research teams (one in the United States and the other in Australia) using sophisticated telescopes on the ground and in space added another piece to the cosmological puzzle. Among other things, they found that not only was the universe expanding, in fact the expansion was accelerating. This new discovery sets the stage for the next exploration to determine if this acceleration is being driven by the enigma known as "dark energy". For their discovery, these three men received the 2011 Nobel Prize: Sal Perlmutter (a professor at my alma mater, U.C., Berkeley); Adam Riess (from Harvard University); and Brian Schmidt (in Australian National University). So the cosmological saga continues.

Were he alive, Kepler would be pleased with all these advances. He has acquired a popular image as an icon of scientific modernity and a man before his time. Science's Carl Sagan described him as *"The first astrophysicist and the last scientific astrologer."* [10]

EVOLUTION: DARWIN & LEAKEY

Did we really start as monkeys like Darwin implied? Charles Darwin was an English naturalist whose theory of evolution became the foundation of modern anthropology. He was not a genius like some of his scientific predecessors such as Copernicus, Galileo, or Newton, but instead was a "prodder" who did years of basic "grunt-work" research before finally pulling his ideas together and advancing his theory of natural selection (and evolution). He also was not an ego-driven self-promoter like some of his predecessors. Arriving at his theoretical conclusions the same time as others, particularly a fellow Englishman, Alfred Russel Wallace, he shared his success with Wallace in a joint presentation.

While he had become educated in geology and biology early in life, it was a five-year voyage on the British ship *HMS Beagle* that provided the opportunity to collect the material from around the world that became the basis in support of the theories he later developed. This opportunity came as a by-product of the Beagle voyage whose purpose was to chart the coastline of South America. He was not directly engaged in the charting aspect but was only to be a "suitable (if unfinished) gentleman naturalist for a self-funded place with Captain Robert Fitzroy, more as a companion than a mere collector." [11]

Charles spent most of his time ashore investigating the biology and geology and making natural history collections while the Beagle crew surveyed and charted the coastlines. His notes and collections included fossil bones of extinct mammals in the cliffs beside modern seashells in the Atlantic Patagonia region (now Argentina), and several years later in the Galapagos Islands off the coast of present day Chile, where he conducted extensive studies of the differences among finches on the various islands.

Darwin formulated his theory in private, and it was not until two decades later when he was 50 years of age that he finally gave it full public expression in *On the Origin of Species* (1859), a book that has deeply influenced modern Western society. [12] His theory on Natural Selection as stated in the introduction to the book is as follows:

> *"As many more individuals of each species are born than can possibly survive; and as, consequently, there is a frequently recurring struggle for existence, it follows that any being, if it vary however slightly in any manner profitable to itself, under the complex and sometimes varying conditions of life, will have a better chance of surviving and thus be naturally selected. From the strong principle of inheritance, any selected variety will tend to propagate it new and modified form."* [13]

He put forward a strong case for common descent, but avoided the use of the term "evolution" because of the controversy that existed in the religious society of England at that time. However by the end of his book, he slipped in the word "evolved", and concluded with the following:

> *"There is grandeur in this view of life, with its several powers, having been originally breathed into a few forms or into one; and that, whilst, this planet has gone cycling on according to the fixed law of gravity, from so simple a beginning endless forms most beautiful and most wonderful have been and are being evolved."* [14]

Darwin remained a bachelor until he was 30 years of age (a smart man -- the same age as when I married). As he jotted down research notes about animal breeding he also scrawled rambling notes about career and marriage on scraps of paper; one with two columns headed "To Marry" and "Not to Marry". [15] While many confirmed bachelors (such as I was) are apprehensive about the marriage thing, few subject their analysis to such formal scrutiny. Perhaps his dilemma was because the girl he was considering was a first cousin, Emma Wedgwood. In addition to marriage with a first-cousin and the possible offspring consequences from that interrelationship, their religious issues were also complex, particularly in the polarized English social and religious environment. Darwin's father was in the Anglican Church, but the young Charles attended the Unitarian Chapel with his mother. During his lifetime, Darwin became agnostic. Emma had strong Unitarian beliefs. They were married in an Anglican ceremony arranged to suit the Unitarians, and then they immediately departed for a new home elsewhere.

Their religious plight reminds me of my own parents who were married amidst religious differences. My father was Catholic and our family attended the Catholic Mass every third Sunday when a travelling missionary came to Buffalo Gap. The other two Sundays we children went to Methodist Sunday School with Mother. Her parents were Methodist and her father was a deacon in the church and openly anti-Catholic. They were married in her parent's home by a beleaguered priest in a tense atmosphere, and immediately left for their new home a hundred miles away in Camp Crook. Religion was no issue in their new environment because Camp Crook was an isolated region with mostly infidel cowboys and neither a Catholic or Methodist Church. God must have blessed their marriage to create such a forgiving environment.

Why was Darwin's work so important and remains controversial still today? It is because it cuts to the core religious belief of many people. The Book of Genesis of the Bible describes the divine creation of the world in six days and designation of the seventh as the Sabbath. Man and Woman are created in the sixth day to be God's agents over his creation. Darwin's Theory of Evolution directly challenges this account. If a person accepts the theory by Darwin, then it calls into question this story in the Bible.

As could be expected from such a revolutionary theory, the publication of *On the Origin of Species* created a storm of controversy and often ridicule. One review claimed it was a creed of the "Men from Monkeys" idea from earlier fiction. Cartoonist parodied animal ancestry in an old tradition of showing humans with animal traits, and showed Darwin in caricature as an ape. However, by the end of the decade, most scientists agreed that evolution occurred, but only a minority supported Darwin's view that the chief mechanism was natural selection. [16]

This was the decade of the American Civil War, so Americans had other things on their mind.

Darwin's wrote in his autobiography that it was a book for his grandchildren, rather than for publication. It was particularly candid on his dislike of Christian myths of eternal torment. To people who inquired about his religious beliefs; however, he would only say that he was an agnostic. [17]

An affable country gentleman, Darwin shocked religious Victorian society by suggesting that animals and humans shared a common ancestry. However, his nonreligious biology appealed to the rising class of professional scientists, and by the time of his death evolutionary imagery had spread through all of science, literature, and politics. Darwin, himself an agnostic, was accorded the ultimate British accolade of burial in Westminster Abbey, London. [18]

Even though Darwin had considerable research and factual data in support of his theory of evolution, it remained an unproven theory in the minds of many people for decades. A fundamental question was always raised in a challenge to the theory: if man descended from apes, where is the "missing link" between the two? The answer brings us to Louis Leakey and other members of his family with their work as archaeologist in the twentieth-century.

Louis Leakey was a British archaeologist born in 1903 (one year after the birth of my own mother.) His work was important in establishing human evolutionary development in Africa. He was a prime mover in palaeoanthropological inquiry. (This word has 21 letters and is one of the longest in my vocabulary; I've been looking for a place to use it.) He conducted expeditions and made early discoveries; then motivated family members and the next generation to continue the work that gradually provided proof of the theory of evolution. The Leakey's found the "missing link" which provided the proof for the evolution of man.

Louis Leakey was raised by his missionary parents in British East Africa, now Kenya. He grew up, played, and learned to hunt with Africans, learned to speak the language fluently, and was initiated into their ethnic group. As a child he collected things in the local environment, which predisposed him toward a career in archaeology. He later obtained an academic background in the subject at Cambridge University in England. [19]

In one of the fortuitous accidents that often can change history, while cleaning two skeletons he had found, he noticed a similarity to one found by another archaeologist in the Olduvai Gorge in Africa. In

1931 he led an expedition to the Olduvai Gorge. It is a 30 mile long steep-sided ravine in the Great Rift Valley that stretches through Tanzania in Eastern Africa. After many decades of exploration and research, this ravine became an important prehistoric site and is sometimes called "the Cradle of Mankind." [20]

If one is to find the earliest prehistoric man, it has to occur in the landscape in which he lived. Unfortunately that landscape in most places is buried very deep as a result of millions of years of deposits on the crust of the earth and movement of tectonic plates. The Olduvai Gorge in the Great Rift Valley provided a unique opportunity, because as the rift was created with the movement of tectonic plates, it exposed the sides of cliffs that contained the early landscape. The skeletons of early man could then be found by archaeologist virtually lying on the surface.

While there were many anthropologists who made important discoveries during the past century, the members of the extended Leakey family stood above all others in the quantity and importance of what was found. That included Louis and his two wives Fridie, and Mary; their children Richard, Jonathan, and Philip; and numerous offspring continuing to work today in Africa.

So, what have anthropologist discovered in support and confirmation of Darwin's Theory of Evolution? They found that the most common ancestor of all current life on earth is estimated to have lived some 3.5 to 3.8 billion years ago. [21] At some point in time, these early organics had evolved into primates such as monkeys, apes, and gibbons. Pliopithecus is an extinct primate whose fossils were first discovered in 1837 in France and subsequently found elsewhere in Europe. [22] It was the first in a long chain of species of evolution that eventually led to Rhodesian Man, Neanderthal, and Cro-Magnon Man. These were the "missing links" sought by anthropologist such as the Leakey's.

Our species, Homo Sapiens (which means "wise man" or "knowing man"), originated in Africa about 200,000 years ago, reaching full behavioral modernity about 50,000 years ago. [23]

How are anthropologists able to establish these early dates for man? There is an explanation in the endnotes. [24]

When I was a boy living in Buffalo Gap, one of the local families had ape-like looks and lifestyle; so, I've never had any trouble in accepting the theory of evolution. I suspected one of my 5th grade classmates might have been the "missing link".

PSYCHOLOGY AND YOU
Study of the Soul

Psychology is the study of behavior and in a sense may be the oldest science. When a cave man perceived a strange creature in his midst he had to discriminate what this meant, consider his options, motivate himself to do something, and finally take action and launch his spear. Those four stages of behavior: perception, discrimination, motivation, and action were the basics of psychology taught to me in my 1948 college course in Psych 101.

I was an engineering student at U.C. Boulder and buried under technical studies during my first two years, but in my 3rd year finally got to choose an elective. Since I knew absolutely nothing about psychology, that was my choice. A study of behavior -- what does that mean? I remember my first week in class as the professor began to unfold the outline of how behavior was compartmentalized and defined -- perception, discrimination, motivation, and action. I was amazed that my own behavior could be defined in such real terms. I also began to gain an insight into the concept of a science: possession of specialized knowledge based on natural laws.

It is surprising that as basic as behavior is to the human experience, psychology was not recognized as a science for many centuries. In the words of Rodney Dangerfield, "It got no respect." Philosophy and biology emerged as the dominate sciences, and behavior (psychology) was addressed only in a philosophical context.

The word "psychology" comes from the Greek words for soul and the study of. That is an apt description because as we look at the behavior or man, we look into his soul. Psychologists explore such concepts as perception, cognition, attention, emotion, motivation, personality, behavior, and interpersonal relationships.

The Latin word *psychologia* was first used during the renaissance by a Croatian, Marko Marulić, in his book, *Psichiologia de ratione animae humanae*. The earliest known reference in English was by Steven Blankaart in 1694 in *The Physical Dictionary* that refers to "Anatomy, which treats of the Body and Psychology, which treats of the Soul."[25] It was not until 1879 when a German physician, Wilhelm Wundt, introduced psychological discovery into a laboratory setting. He is known as the "Father of Experimental Psychology." He focused on breaking down mental processes into the most basic processes -- a precursor to Psych 101. A few years later, American William James in his book, *Principles of Psychology*, laid the foundation for many of the questions that psychologist would explore for years to come.

During the twentieth century the science took off in many different directions as various theories were advanced to explain behavior, and the profession saw the establishment of experimental psychology, clinical psychology, and psychiatry. Each of these had its own history. The following is a brief review of some of the theories.

Structuralism: Wundt approached psychology by breaking down mental processes into the most basic components.

Functionalism: American William James felt psychology should have a practical value and find how the mind can function to a person's benefit. Others explored the study of memory and developed models of learning and forgetting. The Russian, Ivan Pavlov, discovered a learning process in dogs that was later termed "classical conditioning".

Cognitive: A branch concerned with information and its processing that later constituted a part of the wider "cognitive science".

Psychoanalysis: [26] This was the branch of psychology developed at the turn of the twentieth century by the Austrian physician, Sigmund Freud. It was a method of investigation of the mind and the way one thinks; a systematized set of theories about human behavior; and a form of psychotherapy to treat emotional distress. Freud's theory was largely based on interpretive methods, introspection, and clinical observations. It became well known because it tackled sensational subjects such as sexuality, repression, and the unconscious mind. These were considered taboo subjects at the time and Freud provided a catalyst for them to be openly discussed in polite society. Clinically, Freud pioneered the method of free association and therapeutic interest in dream interpretation.

<u>Behaviorism</u>: In the United States, behaviorism became a dominant school of thought during the 1950s, championed by Skinner. It emphasized the ways in which people might be conditioned to behave in certain ways: behavior determined by associations between stimuli and the pleasure or pain that follows. Research consisted of animal experimentation. Skinner had a philosophical inclination and believed that psychology should emphasize the study of observable behavior.

Skinner did the famous experiments involving pigeons. He placed a series of hungry pigeons in a cage attached to an automatic mechanism that delivered food to the pigeon. He found the pigeons associated the delivery of food with whatever chance actions they had been performing as it was delivered, and they subsequently continued to perform those same actions. [27]

<u>Humanism</u>: The approach sought to glimpse the whole person -- not just the fragmented parts of the personality or cognitive functioning. In 1943, Abraham Maslow posited that humans have a hierarchy of needs, and it makes sense to fulfill the basic needs first (food, water etc.) before higher-order needs can be met. This Maslow theory of hierarchal needs was the prevailing framework being taught in my college classes in the 1950's, and what I still consider as a valid approach to understanding behavior. The hierarchy of needs is portrayed in the shape of a pyramid, with the largest and most fundamental needs at the bottom. These basic level of needs must be met before the individual will focus motivation upon the higher level needs. Here is the hierarchy pyramid, starting at the bottom.

<u>Physiological needs</u>: these are obvious -- air, water, food,

clothing, shelter, and sex -- they are the requirements for survival.

Safety needs: with physical needs satisfied, safety takes precedence and dominates behavior. Safety and Security needs include: personal security, financial security, health and well-being, and a safety net against accidents/illness and their adverse impacts.

Love and belonging: After physiological and safety needs are fulfilled, the third layer of needs is social and involves feelings of belongingness. The need is especially strong in childhood and can over-ride the need for safety. Deficiencies with respect to this aspect of Maslow's hierarchy - due to neglect, shunning, ostracism etc. - can impact an individual's ability to form and maintain emotionally significant relationships, such as friendship and intimacy.

Esteem: all humans have a need to be respected and to have self-esteem and self-respect. Maslow noted two versions of esteem needs, a lower one and a higher one. The lower one is the need for the respect of others, the need for status, recognition, fame, prestige, and attention. The higher one is the need for self-respect, the need for strength, competence, mastery, self-confidence, independence and freedom. The latter one ranks higher because it rests more on inner competence won through experience. Deprivation of these needs can lead to an inferiority complex, weakness and helplessness.

Self-actualization: this level of need pertains to a person realizing their full potential. This is the desire to become everything that one is capable of becoming. As mentioned before, in order to reach this level of need one must first not only achieve the previous needs, physiological, safety, love, and esteem, but master these needs. [28]

<u>Gestalt psychology</u>: is a theory that originated with German psychologist that individuals experience things as unified wholes. Rather than breaking down thoughts and behavior to their smallest element, the Gestalt position maintains that the whole experience is important, and the whole is different than the sum of its parts. [29] What does that mean? When I was in graduate school at U.C., Berkeley, I had two close friends who were PhD candidates in psychology and both indicated they were believers in Gestalt psychology and attempted to explain it to me. To this day I do not have a clear understanding of what it is. "The whole is greater than the sum of the parts" -- what does that imply? I can't argue with that because I'm not sure of the implications for psychology.

<u>Existentialism</u>: This emphasized the human themes of death, free will, and meaning, suggesting that meaning can be shaped by patterns of life, and encouraged by acceptance of free will and authentic regard for death. It grew out of reflections garnered by Holocaust survivors during their internment.

<u>Cognitivism</u>: This studies mental processes including how people think, perceive, remember, and learn. The branch of psychology is related to the disciplines including neuroscience, philosophy, and linguistics.

<u>Bio-psycho-social</u>: is an integrated perspective toward understanding how behavior is affected by interrelated biological, psychological, and social factors.

To some extent, I believe each school of psychology decides to put their own unique "spin" on a subject with their name on it to reflect what they see as their unique insight, or perhaps for purposes of ego or prestige. When I was an executive in the bottle business, we always tried to do the same as we positioned our glass container in the market place against other competitive forms of packaging. Perhaps you remember our slogan: "No Deposit, No Return"?

Psychology can also be categorized along other types of activities: biological, clinical, psychiatry, comparative psychology (animals), educational psychology, and industrial psychology -- important to me career wise since I worked in Industry. Let's briefly look at each.

<u>Biological psychology</u>: is the application of the principles of biology to

the study of physiological, genetic, and developmental mechanisms of behavior. It typically investigates at the level of nerves, neurotransmitters, brain circuitry and the basic biological processes that underlie normal and abnormal behavior. Most experiments involve non-human animals (such as rats and primates) that have implications for understanding human pathology. With the development of new non-invasive analytical tools in recent years that can look at activity within the brain, this area of investigation is highly active.

Clinical psychology: [30] This is one of the more important aspects of this science. Clinical psychology includes application of psychology for understanding, preventing, and relieving distress or dysfunction, and to promote well-being and personal development. Central to its practice are psychological assessment and psychotherapy. The work performed tends to be influenced by various therapeutic approaches, all of which involve a formal relationship between a professional and the client. The therapeutic approaches explore the nature of psychological problems, and encourage new ways of thinking, feeling, or behaving.

Psychiatry: [31] is the medical specialty devoted to the study and treatment of mental disorders. These mental disorders include various affective, behavioral, cognitive and perceptual abnormalities. A medical doctor specializing in psychiatry is a psychiatrist.

Psychiatric assessment typically starts with a mental status examination and the compilation of a case history. Psychological tests and physical examinations may be conducted. Mental disorders are diagnosed in accordance with criteria listed in diagnostic manuals such as the Diagnostic and Statistical Manual of Mental Disorders (DSM), published by the American Psychiatric Association. Psychiatric treatment may use medication, psychotherapy and a wide range of other techniques such as transcranial magnetic stimulation. Research and treatment within psychiatry as a whole are conducted on an interdisciplinary basis, sourcing an array of sub-specialties and theoretical approaches.

The trained psychiatrist, who has completed medical school and a psychiatric residency, commonly employs medical treatments in addition to psychotherapy. Electroshock therapy continues to be used for severe depressions and certain forms of psychosis. The medical technique that is by far the most widely used is drug therapy. The advent in the 1950s of psychotropic (mind altering drugs) revolutionized treatment of the mental patient. Like the other medical techniques, drug therapy has sometimes been abused; however, it can enhance a patient's outlook for recovery and return to the community. [32]

Comparative psychology: refers to the study of the behavior of animals other than human beings. Although the field of psychology is primarily concerned with humans, the behavior and mental processes of animals is also an important part of psychological research. This is as a subject in its own right, or as a way of gaining an insight into human psychology.

Educational psychology: is the study of how humans learn in educational settings, the effectiveness of educational interventions, the psychology of teaching, and the social psychology of schools as organizations. Educational psychology is often included in teacher education programs.

Industrial psychology: applies psychological concepts and methods to optimize human potential in the workplace. This is a discipline that I studied for my master degree in industrial engineering, and after a forty-five year career in the industrial environment I can claim some expertise. Early in my career as an industrial engineer, I worked in a glass container factory with 2000 hourly employees who worked around the clock 24 hours per day, 7 days a week, who produced over two million bottles each day. It was a complex social maze and psychology of the worker was at the core of many problems and solutions. I became corporate manager of Quality Assurance for 25 factories, and then later in the international arena provided on-site expertise to 50 factories located in 35 oversea countries. While the social and political environment varied considerably from place to place, the psychology of the worker was predictably similar in Waco, Texas; Lyon, France; Jakarta, Indonesia; and Oakland, California.

That is a brief review of the subject of psychology. Criticisms of psychological research often come from perceptions that it is a "soft" science. Its parameters are often subjective, so it does lack agreement on overarching theory found in mature sciences such as chemistry and physics. Because some areas of psychology rely on research methods such as surveys and questionnaires, critics have asserted that psychology is not an objective science. Other phenomena that psychologists are interested in, such as personality, thinking, and emotion, cannot be directly measured and are often inferred from subjective self-reports, which may be problematic. Some psychologists (and also in other professions) confuse statistical significance with practical importance. Sometimes the debate comes from within psychology, for example between laboratory-oriented researchers and practitioners such as clinicians. In recent years, there has been increasing debate about the nature of therapeutic effectiveness.

This science may contain a systemic bias. In 1959, a statistical study examined the results of psychological studies and discovered that 97% of them supported their initial hypotheses, implying a possible publication bias. Another study found that 91.5% of psychiatry/psychology studies confirmed the effects they were looking for, which was around five times more often than in most other sciences, and argued that this is because researchers in "softer" sciences have fewer constraints to their conscious and unconscious biases. [33]

Licensing requirements for American psychologists are regulated by their professional organization, the American Psychological Association (APA). The APA stipulates that in most divisions the bearer of the title "psychologist" must have a doctoral degree from an accredited institution. Principal employment settings include educational institutions, hospitals, prisons, business and industry, military establishments, and private practice. Many psychologists pursue a combination of private practice or consulting, research, and teaching. [34]

Despite the criticisms and shortcomings of the science, primarily due to its inherent subjective nature, psychology remains one of the most important sciences we have today in our understanding of ourselves and our world.

THE SCIENCE OF MOTION

Motion has several different meanings, but I will talk about movement – that related to the moon or to an apple falling from a tree. This chapter is somewhat out of historical chronology for good reason. It is rather complex technically and may be a turn-off for some readers as a beginning chapter. Perhaps you should take a motion pill. Have no fear -- I will simplify the subject as much as possible while still leaving some substance. The science of motion is important to understanding our world; how else would we appreciate what man was able to do in traveling to the moon and back?

The story begins with Galileo, a colorful Renaissance Italian in Pisa and Florence who probably had a bottle Chianti with his pasta. He lived a "21st century Hollywood style" personal life fathering three children out of wedlock with the same woman that he never did marry. Because of the daughters' illegitimate birth, he considered them unmarriageable and sent them to the convent, where they eventually became nuns.

He is now is considered the Father of Modern Science. According to Stephen Hawking, "Galileo, perhaps more than any other single person was responsible for the birth of modern science. Starting with him, physics has been extremely successful in discovering principles and laws that are in close agreement with the observed facts of the world." [35]

It is amazing that Galileo was successful in scientific endeavors, given the harsh religious environment in which he lived. He was a pious Roman Catholic, but was persecuted by the Church. During his lifetime he was in conflict with the Church for many of his theories, was investigated by the Roman Inquisition in 1615, found guilty of heresy, forced to recant, and spent the latter part of his life under house arrest.

The genius of Galileo was the wide range of his scientific endeavors that included fundamental science: kinematics of motion, astronomy, and mathematics as well as practical applied science (tides, strength of materials, and optics). Galileo made major improvements to the crude telescope that existed, then using his powerful new telescope, he discovered the four satellites (moons) orbiting Jupiter. [36]

My first college degree was physics, which I studied at the University of Colorado (B.S. Eng. Physics, 1951), and I found the science of motion difficult to digest. It takes considerable motivation to undergo the intense study required to pursue a degree in physics, which must come from within the student, because few physics professors are gifted teachers. They know physics and mathematics, but are often deficient in the ability to effectively explain it to others, particularly motion and electricity. They spend most their time facing the blackboard, talking to the formulas they are chalking. [37]

It was with the concept of motion where Galileo made his greatest contributions. If he ever had an epiphany, it came at the Leaning Tower of Pisa. Prior to his time it was held that a heavier object would fall faster than a lighter one in direct proportion to their weight. As a boy, I also had this intuition and agreed with Aristotle who advanced this theory; I challenged my high school physics teacher who taught otherwise. Galileo inherited this Aristotelian concept that was the theory at his time, but set out to investigate and prove otherwise. He proposed that a falling body would fall with uniform acceleration (in a vacuum). We have all heard about his experiment of dropping two balls of the same size and shape but with different masses from the Leaning Tower of Pisa. (Why the tower in Pisa? Because it was just down the street from where he was raised.) The two balls hit the ground at the same time, demonstrating that their time of descent was independent of their mass (weight). If the matter stopped there, it would not have been of great consequence, but Galileo carried the experiment into mathematics. He derived the correct law for the distance travelled during a uniform acceleration starting from rest -- namely, that it is proportional to the square of the elapsed time, or in mathematical terms ($d \sim t^2$). My apologies for use of this mathematical lingo but it is a habit I picked up from my hours at the blackboard in college.

Galileo put forward the basic principle that the laws of physics are the same in any system that is moving at a constant speed in a straight line, regardless of its particular speed or direction. Hence, *there is no absolute motion or absolute rest anywhere in the universe.* This principle provided the framework for Newton's *Laws of Motion* and is central to Einstein's *Theory of Relativity.* [38]

This brings me to Isaac Newton who climbed on the shoulders of Galileo and carried science to the next level. He was born in the same year that Galileo died, 1642. (A few years after the Pilgrims celebrated the first Thanksgiving in America.) Newton was raised a small town boy (like me) born in a hamlet in the north of England. His father died before he was born and his mother remarried. Newton never accepted the marriage and threatened to burn his mother and step-father "and the house over them." He was once engaged, but never married, being highly engrossed in his studies and work. Apparently, he had various social issues like Galileo. [39]

Many consider Newton as the greatest scientist who ever lived. He was a physicist, mathematician, astronomer, philosopher, and theologian. His famous book, *Philosophiae Naturalis Principia Mathematica,* published in 1687, laid the foundation for much of classical mechanics. If I received a dollar for every time I heard his name in my physics classroom, I could have supported the cost of my college tuition. He pervades scientific literature and is still today a major factor in the modern classroom.

What did he accomplish? Newton described gravitation and the *Laws of Motion*; explained planetary motion with the *Theory of Gravity*; built the first practical reflecting telescope; developed the theory of prisms forming the visible spectrum; and studied the speed of sound. In mathematics he shares credit for the development of differential and integral calculus, demonstrated the binominal theorem, developed a method for approximating the roots of a function, and contributed to the study of power series. [40]

I had the study of all these things on my plate during my college career -- no wonder I spent my evenings at the library and had such a lousy social life. Since that was sixty years ago, I do not want to be tested on any of them; however, I do still have some understanding of science.

I will make only a brief mention of gravity in connection with Newton, because I will deal with the subject in more detail in a later chapter on Einstein. In physics, an inverse-square law states that a physical quantity is inversely proportional to the square of the distance from the source of that quantity. This applies to a number of things such as radiant heat where the intensity of the heat radiated by a hot body is governed by the law. We all know that the closer you approach a bonfire, the radiant heat from the fire increases and gets increasingly hotter as you approach closer – an application of the inverse-square law. Now let's see how it applies to gravity.

Newton himself told the story that by watching an apple fall from a tree he got his idea for the theory of gravitation. He wondered why it fell in the direction of the earth. He then postulated that it fell because of an attraction between the apple and the ground -- which he called gravity. His law states:

> "The gravitational attraction force between to point masses is directly proportional to the product of their masses and inversely proportional to the square of their separation distance." [41]

Newton's Law provides an accurate approximation for most situations; however, it was found to be inadequate to explain some things such as the erratic movement of the planet Mercury, which increased its speed during its orbit around the sun. Einstein in his Theory of Relativity gave a different explanation for gravity, explaining it was not attraction, but a consequence of the curvature of space-time that governed the motion of objects. I will have a more extensive explanation in a later chapter on Einstein. [42]

The biggest contribution of Newton to science was his formulation of the three *Laws of Motion*. In this regard, he stood on the shoulders of Galileo who had put forth the principle that the laws of physics are the same in any system that is moving at a constant speed in a straight line, regardless of its particular speed or direction. Newton's First Law used this as a starting point and then extends it further. His First Law:

"*An object at rest tends to stay at rest and an object in motion tends to stay in uniform motion unless acted upon by a net external force.*"

An example to illustrate this First Law would be a satellite in outer space once all its rocket motors ran out of fuel. Whatever speed it had at that time it would continue forever. Here on earth it is more difficult to find illustrations because every object chosen would be acted upon by other forces such as friction or wind.

Newton's Second Law:

"An applied force F on an object equals the rate of change of its momentum P with time."

This can be simplified to "force is equal to the mass of an object times its acceleration", or ($F = dP/dt$). Opps, I automatically slipped into a mathematical formula. My apologies. An example to illustrate this second law is in baseball: the distance a batter can hit the ball depends on the weight of the ball, which is why the weight of all baseballs is carefully controlled to make them exactly the same. This Second Law is also the basis the highway police use in calculating momentum to estimate the speed an auto was travelling at the time of a crash.

Newton's Third Law:

"For every action there is an equal and opposite reaction."

An example is two ice skaters pushing against each other and sliding apart in opposite directions. Another example is the recoil of a firearm in which the force propelling the bullet out the barrel is exerted equally back into the gun and felt by the shooter. [43]

Newton was a social maverick in his day. Although he was highly religious, he was an unorthodox Christian and disputed the existence of the Trinity. He considered himself a theologian, and his writings included more about his literal interpretation of the Bible and occult studies than on science and mathematics. [44]

So, why were Newton and his three laws so important? It was for a number of reasons.

First, these became the foundation of much of modern physics. In my college classes of sixty years ago, the starting point for many lectures began when the professor chalked a Newton formula on the blackboard.

Second, the profound reach and breath of these three laws placed science on a sound philosophical basis.

Third, it gave science greater credibility as a follow-up to Galileo (the Founder of Modern Science) who was seen by many as only a tinker and was tarnished by the religious trials of the church inquisitions.

Fourth, although Newton was highly religious, he was an unorthodox

Christian and gave comfort to those who rejected the rigid theology of the Roman Catholic and English Anglican churches.

Fifth, in the international science competition of that day that was highly nationalistic, Newton gave the English scientific community a national hero to match against Galileo of Italy, Copernicus of Poland, Kepler of Germany, and others on the continent. He became the ultimate English scientific hero.

Despite his off-beat religious views that were in conflict those of the Church of England, Newton was knighted by the Queen and is buried in Westminster.

MATHEMATICS

No wonder I've always struggled with mathematics -- recently I learned why. I never knew what mathematics was. Now I know what I did not know, and that few others knew either. *"It has no generally accepted definition."* [45]

Aristotle (in Ancient Greece), Bertrand Russel (1903), and others through the centuries have not agreed on what it was. The famous philosopher Henri Poincare said: *"Mathematics is the art of giving the same name to different things."* [46] Today, in the era of huge computers, the mathematician is a kind of zookeeper, making sure precious computers are fed and watered. [47]

While mathematics is not well defined and a fixed body of knowledge, it does, however, pervade every aspect of modern life. It is a collection of tools that help us understand parts of the real world. The renaissance scientist, Galileo Galilei, said:

> *"The universe cannot be read until we have learned the language and become familiar with the characters in which it is written. It is written in mathematical language, and the letters are triangles, circles and other geometrical figures, without which means it is humanly impossible to comprehend a single word. Without these, one is wandering about in a dark labyrinth."*

Albert Einstein stated: *"as far as the laws of mathematics refer to reality, they are not certain; and as far as they are certain, they do not refer to reality".*[48] Does that statement by Einstein clarify everything? I think not!

The earliest use of mathematics was in trading, measurement, and the recording of time. Complex mathematics did not appear until around 3000 BC, when the Babylonians and Egyptians began using arithmetic, algebra and geometry for taxation, construction, and astronomy. The Ancient Greeks began using it in 600 BC.[49] Discoveries continue to be made today and many items are added to the *Mathematical Review* database each year. For example, the physicist Richard Feynman invented *"The path integral formulation of quantum mechanics using a combination of mathematical reasoning and physical insight."* (Wow! Think of that – I don't know what he is talking about?)

Since the field is too broad to attempt a summary, I will confine my observations to my own excursions that started at the blackboard in the first grade in Buffalo Gap, through my high school education in Edgemont that included algebra, geometry, and trigonometry, and on to my education in physics at the University of Colorado where the entire load of calculus and advanced mathematics was dumped on my plate.

I found algebra to be the most difficult mathematics to master. It is the study of structure, starting with numbers. It is difficult because people are subjected to it when they are young with little background, and structure is a complex philosophical concept. I recently picked up my grandson's middle school algebra textbook and quickly became lost. My impression was that it was probably a much better approach to the subject than the blackboard exercises I received in that red sandstone school in Buffalo Gap, but thankfully I no longer will be tested on the subject.

I loved geometry and trigonometry that deal with spatial relationships and the properties of space. I can get my arms around the physical relationships of circles, triangles, and squares. (In college I once dated a girl who thought I was a "square".)

Calculus focuses on limits, functions, derivatives, integrals, and infinite series and constitutes a major part of our modern mathematics education. It is the study of change in the same way that geometry is the study of shape and algebra is the study of operations and solving equations. A course in calculus is a gateway to other mathematics, and can solve problems for which algebra alone is insufficient. I found calculus easy. It is at the heart of a physics education in college because all phenomena are reduced to a mathematical treatment; class usually began when the professor wrote a calculus formula on the blackboard, and began his dissertation about what phenomena it led to.

In our modern day, we have additional fields of mathematics that are discussed in the following paragraphs.

<u>Statistics</u> is the study of the collection, organization, analysis, and interpretation of data. It deals with the use of data collection in surveys and experiments.

The <u>binary numeral system</u> (or base-2 number system) represents values using two symbols, 0 and 1. The binary system is used internally by almost all modern computers. I will have more discussion on this system of mathematics in my chapter on computers.

<u>Chaos theory</u> is a field with applications in several disciplines including physics, engineering, economics, biology, and philosophy. It studies the behavior of systems that are highly sensitive to initial conditions. Small differences in initial conditions can yield widely diverging outcomes for some phenomena, rendering long-term prediction difficult. Chaotic behavior can be observed in many natural systems, such as weather, where the explanation may be sought through analysis of a chaotic mathematical model.

<u>Quantum mechanics</u> is a branch of physics providing a mathematical description of the dual particle-wave like behavior and interactions of energy and matter. It departs from classical mechanics at the atomic and subatomic scales. The word "quantum" derives because some physical quantities can change only by discrete amounts, or quanta. Quantum mechanics has now branched out into almost every aspect of 20th century physics. I will discuss it again in the later chapter about scientific research.

<u>Game theory</u> is a method for analyzing calculated circumstances, such as in games, where a person's success is based upon the choices of others. It is the study of mathematical models of conflict and cooperation between intelligent rational decision-makers. The theory is mainly used in economics, political science, and psychology.

Let me return to where I started at the beginning of the chapter. The academic definition of mathematics is debatable, but all would agree that over the centuries it has been one of the most important tools to help us understand the world around us. It also serves another purpose; the homework assignment in fifth-grade algebra creates one of the strongest bonds between a parent and the child who is struggling for a passing grade. "Dad. Can you please help me with my math homework? I'm stuck."

HEREDITY AND THE GENOME

What began a century and half ago with a monk raising peas in his abbey garden eventually led to one of the most important discoveries of the 20th century -- understanding the genome.

Heredity dominates many family conversations with discussions involving who in the family resembles which ancestor. That seems to be typical of my wife's Italian side of the family where conversation (among women behind closed doors) will insinuate that Aunt Millie's son, Guissippi, bears no resemblance to her husband -- "oh, she always had a roaming eye, don't you know!" It is unlikely these discussions contain any mention of Friar Gregor Mendel and his pea garden where the heredity thing started.

In my high school biology class I never understood Mendel's pea experiments and what they revealed about heredity, so I took myself back to school on the subject. As a friar in the Augustan abbey of St. Thomas located in the city of Brno, Czech Republic (1856-63), Mendel had a lot of time on his hands since he was not encumbered with a wife and kids. So he tended his pea garden on the abbey grounds and kept elaborate notes concerning the harvest.

Mendel spent eight growing seasons cultivating as many as 10,000 pea plants and meticulously counting some 40,000 blossoms and 300,000 peas. He never described his motivation for his breeding experiments, but some speculate that he was investigating a theory that hybridization created new species. Growing up on his family's farm, he had tended fruit trees. Years later, his university studies included physics and mathematics -- disciplines that imparted proper scientific rigor.

Mendel's genius lay in his decision to study the inheritance patterns of specific plant traits (such as round or wrinkled seeds) separately from one another. His research yielded two significant principles. The first law of inheritance states that traits are determined by a pair of "factors" known today as alleles or paired genes. One is dominant, the other recessive -- and the offspring receive a random allele from each parent. The second law of independence states that allele pairs for each trait occur independently of one another. [50]

Efforts by other naturalists to duplicate Mendel's finding using other species often failed. Today we know the reason: most traits are determined by several gene pairs acting in tandem. Relatively few traits, such as the shape of a pea seed, are determined by just one pair of alleles. And we now know that some genes are passed down in groups. [51]

Even though he published his results, the importance of his finding was not recognized in his lifetime. It was not until fifty years later in the early 20th century that scientists recovered Mendel's work and recognized its significance including the implications for evolutionary biology. [52]

Now let us move ahead a century after Mendel to the 1950's and the Cavendish Laboratory in England where James Watson and Francis Crick were conducting research to explore the structure of genetic material called DNA (Deoxyribose Nucleic Acid). That quest for discovery occupied some of the foremost scientist at the time including the great scientist, Linus Pauling in California, and it had become a discovery race to see who could first identify the bio-chemical structure. This structure, whatever it was, would surely be heralded as one of the most important discoveries of the 20th century, and Pauling had given it priority for several years. Watson and Crick won the race and in March 1953 announced the double helix structure of DNA. For this accomplishment they were awarded the 1962 Nobel Prize in Physiology or Medicine "*for their discoveries concerning the molecular structure of nucleic acids and its significance for information transfer in living material.*" [53]

With this discovery, the missing link in the heredity puzzle had been found by Watson and Crick. Before we go further we need a brief side trip to fit genetic material into context within the human cell.

The cell is the basic unit of all known living organisms. It is the smallest unit of life that is classified as a living thing, and is often called the building block of life. Humans contain about 10 trillion cells. The cell was discovered by Robert Hooke in 1665, so we've known about it for a long time. All organisms are composed of cells, and they contain the hereditary information for transmitting information to the next generation. A cell has three regions: on the outside are structures that facilitate communication between cells; enclosing the cell is the envelope; and inside the cell is the region that contains the gene. [54]

Genetics is the study of genes. Genes are how living organisms inherit features from their ancestors. They are made from a long molecule called DNA, which is copied and inherited across generations. Each unique form of a gene is called an allele. For example, on allele for the gene of hair color could instruct the body to produce a lot of black hair pigment. [55]

Genes are inherited from the each parent. When people reproduce, their sperm and eggs contain copies of their genes. An egg joins with a sperm to create a new life, and this gives a child a complete set of genes. [56]

Many traits are inherited in a complicated way. This can happen when there are several genes involved, each contributing a small part to the end result. Tall people tend to have tall children because their children get a package of many alleles that each contributes to how much they grow; however, there are not clear groups of "short people" and "tall people". Height is determined by the interaction of 200 gene regions. [57]

Inheritance can also be complicated when the trait depends on the interaction between genetics and the environment. This is quite common, for example, if a child does not eat enough nutritious food it will not change traits like eye color, but it could stunt their growth. Some diseases are hereditary and run in families; others, such as infectious diseases, are caused by the environment. Some come from a combination of genes and the environment.

Genetic disorders are diseases that are caused by a single allele of a gene and are inherited in families. These include Huntington's disease and Cystic fibrosis. Cystic fibrosis, for example, is caused by mutations in a single gene called CFTR and is inherited as a recessive trait. The genes a person gets from their parents only change their risk of getting a disease. Most diseases are inherited in a complex way with multiple genes involved or coming both from genes and the environment. The risk of breast cancer is 50 times higher in the families most at risk, and the variation is due to a large number of alleles, each changing the risk a little bit. Although the risk for breast cancer is genetic, it is also increased by being overweight, drinking a lot of alcohol and not exercising. So a woman's risk of breast cancer comes from a large number of alleles interacting with her environment, and is very hard to predict. [58]

The structure of this living material inside the human cell, the gene that gives us life from one generation to the next, is both chemically simple yet conceptually complex. The DNA is a long molecule that looks like a twisted ladder (what Watson called the "double helix"). It is made of four types of bio-chemical units and the sequence of these units carries the information about the gene structure. The nucleotides form the rungs of the DNA ladder. There are four types of nucleotides (A, T, G, and C) and it is this sequence that carries the genetic information. The four chemical base molecules found in DNA are adenine (abbreviated A), cytosine (C), guanine (G) and thymine (T). These four bases are attached to the two sides of the double helix to form the complete nucleotide. The genes are contained in the chromosome. [59]

The chromosome is a package for carrying DNA in the cells. The complete set of genes in a particular organism is called the genome. [60]

The Human Genome Project was organized to map the human genome. Other genome projects include mouse, rice, the plant Arabidopsis thaliana, the puffer fish, and bacteria like E. coli. In 1976, a scientist at the University of Ghent (Belgium) was the first to establish the complete nucleotide sequence of a genome. The development of new technologies has made it dramatically easier and cheaper to do sequencing, and the number of complete genome sequences is growing rapidly. These open up the prospect of personal genome sequencing as an important diagnostic tool. A major step toward that goal was the completion of the decipherment of the full genome of DNA pioneer James D. Watson in 2007. Why Watson -- rank has its privilege?

Perhaps no other discovery in the 20th century has had a greater impact than our knowledge of the genome and its relevance in so many ways. The first application occurred in criminology, both in establishing the evidence to prove commission of a crime and to prove the innocence of people wrongly accused. It is now almost routine to collect a sample from a crime scene and compare it to the genome of the person suspected of a crime. For example, it is the crucial evidence to prove rape. There are many examples of men who have spent years in prison being freed after their genome is compared to evidence collected years ago (such as the panties of a girl raped), to find they were wrongly accused as a result of faulty eye-witness accounts or other evidence. However, a man's sample of semen on a girl's panties almost always will lead to a certain conviction of rape, along with other consistent evidence.

Another major use is in the study of disease. Using the genome and comparing it to samples from others, the medical profession is rapidly identifying the susceptibility of an individual for a given disease. Examples include cystic fibrosis, Huntington disease, and breast cancer that have been traced to specific sites in the genome. In a few years it is likely that numerous other disease relationships will be revealed by medical science.

Another remarkable use of the genome is in anthropology and revealing world migration patterns. Using the genome of people from various continents and regions of the world, scientists have been able to trace the route of early peoples from Africa as they dispersed through Europe, then Asia, the Americas, etc, around the world. After decades of speculation about such patterns, the genome has now provided us with concrete evidence of how our world was colonized.

Yet another use is in genetic engineering. Since traits come from the genes in a cell, putting a new piece of DNA into a person's or plant's cell can produce a new trait. For example, crop plants can be given a gene from an arctic fish, so they produce an antifreeze protein in the leaves, which can help prevent frost damage. This kind of technology is also being developed to treat people with genetic disorders in an experimental medical technique called gene therapy. Gene therapy works by trying to replace the allele that causes a disease with an allele that will work properly.[61]

In a few years, perhaps, most people in America will have personal knowledge of their genome. Whether this will be a blessing or a curse remains to be seen. I already personally struggle with enough medical problems that I'm not sure I want to know what additional potential problems await me around the next corner. However, I guess it will be okay if they have a pill that I can swallow or gene therapy to head it off before I get there.

CHEMISTRY: MARIE CURIE & LINUS PAULING

Marie Curie was awarded a Nobel Prize for both chemistry and physics and can be called either a chemist or physicists -- perhaps we should just call her a scientist. When I was a youth during the 1930's, she was very popular and still living. A Hollywood movie, *MADAME CURIE*, starring Greer Garson with Walter Pidgeon in 1943 was nominated for an Oscar. Madame Curie is recognized as the most famous female scientist in history. She was much beloved in America when she visited here, also in France where she did her work, and is a national hero in her native Poland.

Marie Sklodowska - Curie was born in 1867 in Warsaw, which was then the Kingdom of Poland. At the age of 24, she moved to Paris where she earned higher degrees and began scientific work. Her achievements included a theory of radioactivity (a term that she coined), techniques for isolating radioactive isotopes, and the discovery of two elements, polonium and radium. Under her direction, the world's first studies were conducted in the treatment of disease using radioactive isotopes. She founded the Curie Institutes in Paris and Warsaw, which remain major centers of medical research today. [62]

In 1903 she shared the Nobel Prize in physics with her husband,

Pierre Curie, and in 1911 she was the sole winner of the Nobel Prize in chemistry. She was the first woman to win a Nobel Prize, the only women to win two, and win in multiple sciences. When asked to comment on the difficulty of being a woman and a scientist, this was her reply:

> "I have frequently been questioned, especially by women, of how I could reconcile family life with a scientific career. Well, it has not been easy." [63]

Marie Curie was the consummate chemist of her era, but also a wife and mother who helped raise a family. Her daughter, Irene Joliot-Curie, also won the Nobel Prize. Jointly with her husband, she was awarded the Nobel Prize for Chemistry in 1935 for the discovery of artificial radioactivity.

The genesis of chemistry can be traced to alchemy that had been practiced for several millennia in various parts of the world. [64] Egyptians pioneered chemistry 4,000 years ago. Ancient civilizations extracted metal from ores, fermented beer and wine, created pigments for cosmetics, extracted chemicals from plants for medicine, cultured cheese, rendered fat into soap, and melted glass and alloys. The greed for gold led to the discovery of the process for its purification. The early development of these methods is described by the Roman historian, Pliny the Elder, in his Naturalis Historia. [65]

The pioneers of the modern scientific method were Arab and Persian scholars. The science of chemistry was essentially created by Muslims who introduced precise observation, controlled experiment, and careful records. They chemically analyzed innumerable substances, composed lapidaries, distinguished alkalis and acids, and manufactured hundreds of drugs. The later emergence of chemistry in Europe was primarily due to the plague during the Dark Ages that gave rise to a need for medicines. It was thought that a universal medicine could cure all diseases, but was never found.

Chemistry is the science of matter, especially its chemical reactions, composition, structure, and properties. It starts with the study of elementary particles: atoms, molecules, metals, crystals and other matter. Chemical reactions are a result of interaction between different substances and are a transformation of one substance into something else.

(If you find this chemical jargon somewhat dull, consider my plight during an afternoon chemistry class as I made eyes at an attractive burnet and wondered how to meet her after class? She was taking copious notes; so, perhaps a good pitch would be to suggest we might share notes since my pencil broke during the lecture. The professor is droning on and on -- equation this -- equation that! Oops! He just announced a quiz.)

Lavoisier is considered the "Father of Modern Chemistry". He developed the theory of Conservation of Mass in 1783. Lavoisier established the use of the chemical balance, developed a system of chemical nomenclature, and made contributions to the modern metric system. He also worked to translate the archaic language of chemistry into something that could be easily understood by the largely uneducated masses of his day. He lived in the era shortly after the American Revolutionary War when the federal constitution was being written for the newly formed United States of America, and George Washington was to become our first President. Modern chemistry and our country came into being at about the same time. Talk about good chemistry!

The discovery of the chemical elements culminated in the discovery of the periodic table by Dmitri Mendeleev in 1869. The periodic table is a display of the chemical elements organized on the basis of their properties; elements are listed in rows to keep those with similar properties included together. Because the periodic table accurately predicts the properties of various elements, its use is widespread within chemistry and other sciences, providing a useful framework for analyzing chemical behavior. [66]

The giant in chemistry after Marie Curie was Linus Pauling, who came a generation later. He was born in 1901 in the same hospital in Portland, Oregon, where two of our children, Lorie and Deke, were born six decades later. He was the most influential chemist of the 20th century. Pauling is one of the few to have won more than one Nobel Prize; awarded in both chemistry and the peace prize. Pauling wrote papers on the nature of the chemical bond, leading to a famous textbook, *The Nature of the Chemical Bond*, one of the most influential chemistry books ever published. He received the Nobel Prize in Chemistry in 1954: *"For his research into the nature of the chemical bond and its application to the elucidation of the structure of complex substances"*. [67]

In 1958, Pauling and his wife presented the United Nations with the petition signed by more than 11,000 scientists calling for an end to nuclear-weapon testing. Public pressure subsequently led to a moratorium on above-ground nuclear weapons testing. On the day that the treaty went into force, the Nobel Prize Committee awarded Pauling the Nobel Peace Prize with the citation: *"Linus Carl Pauling, who since 1946 has campaigned ceaselessly, not only against nuclear weapons tests, not only against the spread of these armaments, not only against their very use, but against all warfare as a means of solving international conflicts."*

Chemistry introduced us to many of our modern scientific terms such as the following:

- An *atom* is the basic unit of chemistry. It consists of electrons, protons and neutrons.
- A chemical *element* is composed of atoms.
- A *molecule* is a substance that has a unique set of chemical properties.
- An *ion* is a charged atom or molecule that has lost or gained one or more electrons.
- An *acid* produces hydronium ions when dissolved in water.
- A *base* produces hydroxide ions when dissolved in water.
- *Acid strength* is measured by pH, a measurement of the ion concentration in a solution.
- *Ionization* is the ability to lose or gain electrons. Substances that oxidize other substances are said to be oxidizing agents and those that reduce are known as reducing agents.

During chemical reactions, bonds between atoms break apart, resulting in different substances. A chemical reaction can be symbolically depicted through a chemical equation in which the number and kind of atoms on both sides of the equation are equal. Chemistry is typically divided into the following sub-disciplines:

- Analytical chemistry is the analysis of samples to determine their chemical composition and structure.
- Bio-chemistry is the study of things that take place in living organisms, which always involve a carbon makeup.
- Inorganic chemistry is the study of inorganic compounds.
- Materials chemistry is the preparation of substances with a useful function, and is central to *chemical engineer*ing.

- Neurochemistry is the study of chemicals in the nervous system -- a key subject for many physicians.
- Nuclear chemistry is the study of how subatomic particles come together and make nuclei.
- Physical chemistry is the study of the physics of chemical systems. It is a distinct discipline from both chemistry and physics. As an undergraduate, I had to spend three hours on Saturday mornings in the physical-chemistry lab. It helped to have access to a senior classmate's notes so I could get a passing grade -- it was a difficult subject. Absent these notes, you work on into late Saturday afternoon with a substantial penalty to your social life.

The chemical industry represents an important economic activity. The top 50 global chemical producers have annual sales of nearly a trillion US dollars with profit margins approaching 10%. Recent polls have found college graduates in chemistry are some of the highest paid in industry. That is a reflection of the difficulty of the subject and the considerable opportunity to make an impact on very valuable resources.

Good chemists are better paid than good physicists. Fortunately, I opted to become neither and had a successful career in industrial management. However, my underlying knowledge of physics and chemistry were assets that helped along the way.

ELECTRONICS

We will begin with the words "electricity" and "electronics": what is the distinction between the two? Electricity is the flow of electrons in a wire and includes such things as electric motors, generators, and batteries. Electronics begins with an electric current, but has the ability to amplify weak signals using devices such as a vacuum tube (or transistors) to transmit the signals into space. It includes such things as radio, radar, wireless phones, and satellite communications.

The science of electronics started around 1906 with the invention by Lee De Forest of the triode (vacuum tube), which made possible the amplification of weak electric signals. The science was generally referred to as "radio technology" until the time of World War Two, when the development of numerous wartime devices such as radar made the term "electronics" popular. [68]

When I was an engineering physics student at the University of Colorado in the 1940s, a requirement for a student in my major was to design and built a radio receiver. The heart of all electronics at that time was the vacuum tube. It is a device controlling electric current through a vacuum in a sealed container. They are used for the amplification, switching, or similar processing of electrical signals. Vacuum tubes rely on the emission of electrons from a hot filament or hot cathode. Electrons travel from the electrically charged cathode to the anode (or *plate*). A weak signal is placed on an electrode between the cathode and anode that regulates the electric current through the tube, allowing for the amplification of the weak signal to many times greater. These vacuum tubes were critical to the development of electronic technology, which drove radio broadcasting, television, radar, analog and digital computers, and industrial process control. It was the invention of the triode (vacuum tube) and its capability of electronic amplification that made these technologies widespread and practical. [69]

In most applications today, solid-state devices such as transistors and semiconductor devices have replaced tubes. Solid-state devices last longer, are smaller, more efficient, more reliable, and cheaper than tubes. The transistor became the key component in practically all electronics, and many consider it to be one of the greatest inventions of the 20th century.

Although some of the technology was known early in the 18th century, it was Thomas Edison's 1884 investigation that spurred research. However, even Edison did not fully appreciate the potential value of his discoveries. It was not until the early 20th century that the properties of electronic devices were utilized to any extent. In 1906, Lee De Forest developed the triode tube that led to the radio detector and to coast-to-coast telephony. The electronics revolution of the 20th century began with the invention of the triode vacuum tube. [70]

The triode vacuum tube had limitations in its use. It was a fragile device that consumed a lot of power, and was prone to early failure. Scientists looked for ways to accomplish the same control over the flow of electric current by some other means. The breakthrough came in the late 1940s at AT&T Bell Lab with a group of scientists under the team leader, William Shockley. They discovered that when two gold point contacts were applied to a crystal of germanium, a signal was produced with output power greater than the input. In acknowledgement of the accomplishment, Shockley and two coworkers were awarded the 1956 Nobel Prize in physics *"For their researches on semiconductors and their discovery of the transistor effect."* [71]

The first silicon transistor was produced by Texas Instruments in 1954. It was a further improvement over using germanium and the use of silicon became widespread. Many of the later developments involving transistors took place on the San Francisco Peninsula in such cities as Palo Alto, Santa Clara, and San Jose, and that area has since become known worldwide as "Silicon Valley."

The next development in the electronics progression was the microchip. It is an integrated circuit manufactured by the patterned diffusion of trace elements into the surface of a thin substrate of semiconductor material. Additional materials are deposited and patterned to form interconnections between semiconductor devices.

Integrated circuits are used in virtually all electronic equipment today, and have revolutionized the world of electronics. Computers, mobile phones, and other digital appliances are now inextricable parts of the structure of modern societies, made possible by the low cost production of integrated circuits. [72]

Semiconductor devices that could transform the integration of large numbers of tiny transistors into a small chip were an enormous improvement over the manual assembly of circuits using discrete electronics components. "There were two advantages of integrated circuits: cost and performance. Cost is low because the chips, with their components, are printed as a unit by photolithography rather than being assemble one transistor at a time, and less material is used. Performance is high because the components switch quickly and consume little power as a result of the small size and close proximity of the components. As of 2006, typical chip areas ranged upward to a million transistors per square millimeter." [73]

The integrated circuit was developed by Jack Kilby of Texas Instruments in 1958. He won the 2000 Nobel Prize for his part of the invention of the integrated circuit: *"A body of semiconductor material ... wherein all the components of the electronics circuit are completely integrated."*

Noyce, working at Fairchild Semiconductor, also came up with his own idea of an integrated circuit a year later, and his chip solved many practical problems that Kirby's had not. It was made of silicon, whereas Kilby's chip was made of germanium. Fairchild Semiconductor was also home of the first silicon technology with self-aligned gates, which is the basis of all modern computer chips. The technology was developed by an Italian physicist, Federico Faggin in 1968, who later joined Intel in order to develop the first Central Processing Unit (CPU) on one chip.

In the early days of integrated circuits, only a few transistors could be placed on a chip. Over time, millions could be placed on one chip as good design and new design methods were advanced.

While I was on the fringes of an electronics career, I never played a major active role in it. When I was in the U.S. Navy during the Korean War, I was a part-time electronics officer and also the sonar officer for our destroyer. My engineering physics education served me well. When I left the navy three years later, I interviewed for a job in the electronics industry. One of the companies was a small operation in a two-room facility in Palo Alto named Hewlett Packard. I was interviewed by a tall, rangy, farmer-looking guy named Packard. Their only product was a frequency modulator. After a short interview, he explained they wanted a guy with a physics background plus knowledge of electronics to help them develop new products. He took me into the back room where their product was being assembled. He said a new hire would begin by working at the end of the assembly line to perform checks and gain a hands-on knowledge of the product.

During the three years in the navy, my interests had shifted from physics and electronics to more people related activities. I had already been accepted to graduate school in industrial engineering at the University of California, Berkeley, so I decided to begin my new career there. It was a good decision. I received a strong education in industrial engineering that was to serve me well in a long career ahead, and also enjoyed my experience as a teaching assistant on the engineering staff. I became an engineering specialist with people, rather than with electrons.

COMPUTERS

My grandchildren could tell you about the digital computer; they grew up with them. I have witnessed the evolution during my lifetime from primitive analog computers, to the digital computer, and now to a new era of the "cloud".

My introduction to the computer was in Buffalo Gap during the 1930's. Nettie Degnan at the counter of the Degnan grocery store would ring up the bill by pushing the keys on the cash register, and then pull the crank on the side of the store's computer. The amount we owed for groceries would pop-up in a small window on top the cash register. To a little boy, this seemed like magic. That was the analog computer of the 1930's.

At my Dad's bank, an even more elaborate procedure occurred every afternoon after the bank closed to customers. I got to watch as my Dad did the "posting". Placing the ledger sheet for each customer in a machine, he would enter the daily checks and deposits for a customer into a keyboard, then pull a crank that would automatically print the new account balance on the ledger sheet. After doing this for every customer, Dad could then compare the total obtained from all the ledgers to the total for the actual deposits and cash withdrawals. It was a good day when the totals matched; however, when they were either "short" or "long" it meant something was in error, and Dad would curse (Goddamm!), then start the process of manually looking for the problem. When the totals matched, his day's work was done and he would depart for Frenchie's pool hall to join the rummy card game already in process with several local merchants. If he was a winner, he'd bring candy bars home for dessert after dinner. We kid's had a very personal interest in the numbers that came out of that computer.

ANALOG COMPUTERS

All the computers up to that time were analog. They used electrical or mechanical inputs to model the problem being solved. The Renaissance in the 17th century saw the invention of the mechanical calculator, an analog device that could perform arithmetic operations. Some survived for centuries, like the Sumerian abacus designed around 2500 BC. An abacus won a speed competition against a desk calculating machine in 1946. Slide rules were invented in the 1620 (the year the Pilgrims landed at Plymouth Rock). They were carried on five Apollo space missions, including to the moon. [74]

I was introduced to the slide rule in college when all engineering students carried one. It was a manual analog computer used for multiplication, division, and functions such as roots, logarithms, and trigonometry with a standard set of markings essential to performing mathematical calculations. We engineers wore our slide rules in a leather case hanging from our belt like a gun slinger carried his holster, became masters of this manual computer, and it was amazing the speed and accuracy we could obtain. The slide rule became obsolete when digital computing came into being.

During the first half of the 20th century, many scientific needs were met by increasingly sophisticated analog computers that used a mechanical or electrical model of the problem as a basis for computation. However, these were not programmable and generally lacked the versatility and accuracy of modern digital computers. Alan Turing is regarded as the father of modern computer science. In 1936, he provided the concept of computation with the Turing machine, providing a blueprint for the electronic digital computer. Time magazine named Turing one of the 100 most influential people of the 20th century, stating: *"Everyone who taps at a keyboard, opening a spreadsheet or a word-processing program, is working on an incarnation of a Turing machine"*. [75]

A top secret project during World War Two advanced the technology of computers. The code-breaking machine, Colossus, was cracking

German codes by 1943, giving us advance knowledge of German strategies. How does a computer break a secret code? It runs successive trials of every possible combination of words and numbers until it comes up with a version that works. With the huge computers of today, the secret codes of virtually every country in the world have likely been broken and compromised.

During a cruise in 1948 when I was a Midshipman aboard the battleship, the USS Iowa, I was introduced to the Mark 1 Computer, the biggest and most complex analog computer ever used in warfare. I had the unique experience to help man this fire control computer that directed the turrets of 16 inch (diameter) guns, the largest guns ever installed on a battleship. These guns hurled projectiles twenty miles that were 16 inches in diameter and four feet long. The battleship had two turrets on the forecastle and one on the fantail. The Mark 1 computer, deep in the bowels of the ship, did all the necessary calculations and transmitted the electrical signals to the turrets so they could be pointed at the correct azimuth and elevation to reach their target while the ship was moving at a variable speed with pitch and roll and constantly changing directions. Three years later when I was on a destroyer along the North Korean coast, the battleship USS Missouri was further out at sea over the horizon and lobbing missiles over our heads at targets so far inland we could not see or hear the blasts. The Mark 1 computer was being successfully utilized.

During the era of the 1960s, the use of computers was a frustration. As manager in a factory that was dependent on numbers and payroll information, I would often hear "Sorry, no numbers are available today; the computer is down". I did not need to question why -- the answer was always the same. "It is a vacuum tube failure somewhere in the system and we are tracing to find the culprit." Computers were hampered with dependence on the vacuum tube. The first electronic digital computers were normally housed in a large air conditioned room because of the great amount of electrical power they consumed, a consequence of the hot vacuum tubes.

The analog computer era came to an abrupt end in the late 1950's during the time when I was in graduate school at U.C. Berkeley. Their ultimate perfection was found in computers made by two American companies, Marchant Inc. and Friden Inc., in factories in the San Francisco Bay Area. We Berkeley students toured both factories that represented the ultimate manufacturing practices at that time in industrial engineering. These computers -- typewriter size -- sat on desks where they were available for multiplication, division, addition, subtraction, and all the related mathematical functions. They were plugged into an electrical circuit and gave a whirling sound as they quickly cranked out an answer. One of my professors left his job teaching at the university to become the CEO of Friden Calculator Inc. He did not realize that he had inherited a dinosaur; within two years his company's analog product was obsolete -- a victim of the digital revolution.

THE DIGITAL COMPUTER

Two factors came together that made the digital revolution possible: the transistor, and digital mathematics.

The transistor replaced the troublesome vacuum tube, which had problems of heat removal, grid failure, and vacuum leakage. The transistor's advantage was simplicity. It produced virtually no heat, operated without failure, required no vacuum, took little current, operated at low voltage, and gave nearly instant results. All the digital computer needed was a simple switching circuit, and the transistor was ideal for this mission.

Coupled with the development of the transistor was a newly utilized form of mathematics, called Boolean algebra. Again, simplicity was the key. "Digital" comes from the word "digit". Digital computing required the use of only two digits: 0 and 1. The binary numeral system represents numeric values using these two symbols, 0 and 1. Because of its straightforward implementation in digital electronic circuitry using logic gates, the binary system is used internally by almost all modern computers. Utilizing the switching circuits, the digital mathematics could produce an endless array of useful information. A person using a pocket calculator could perform a basic arithmetic operation with a few button presses, but adding together all of the numbers from 1 to 1,000 would take thousands of button presses and a lot of time. On the other hand, a computer may be programmed to do this with just a few simple instructions.

Let me relate a personal story from the mid-1970's when I bought my first computer, since it illustrates the status of digital computers at that time. The Technical Director of an Owens-Illinois affiliate in Europe was visiting me in Toledo and asked me to accompany him to a local store where he heard they were selling the close-out model computer for $8.00. While standing in line with him, I decided that for $8.00 I might as well also become the owner of a computer. I took my new computer home and there it sat virtually unused because of the difficulty in using a computer one could not utilize unless they understood the complex code requirements. My plight was similar to that of the general public before the advent of user-friendly operating systems and computer language. Only experts and early computer "nerds" used the digital computer for anything other than for showoff and conversation.

COMPUTER ARCHITECTURE

What are the main parts of a computer? A general purpose computer has four main components: the processing control unit (CPU), the arithmetic logic unit (or software), the memory, and input/output devices. These parts are interconnected by busses, and inside are thousands of small electrical circuits which can be turned on or off by means of an electronic switch.

CONTROL PROCESSING UNIT (CPU).

The microprocessor is the chip that is the brain of the computer. It carries out the instructions of a computer program to perform the arithmetical operations. Early computers had no operating system. A user loaded a program from a tape to run the program. When the program finished, the computer halted. It was quickly realized by computer designers that this was not satisfactory. For a number of years, the lack of a satisfactory control processing unit retarded the effective development of computers. [76]

In the early 1950's, IBM put together a team to design a computer language because they believed programming would make computers more attractive. They designed the language, which they called FORTRAN (Formula Translation). Programs written in FORTRAN became a staple for the scientific community.

A version of "Basic" code was written for these early personal computers. Large manufacturers such as Hewlett-Packard and Texas Instruments soon moved into the market with mass-produced calculators. As a result, calculators quickly became smaller, more powerful, and cheaper.

SILICON VALLEY

William Shockley, a co-inventor of the transistor, started Shockley Semiconductor Laboratories in 1955 in his hometown of Palo Alto,

California. In 1957 his top researchers left to form Fairchild Semiconductor Corporation. Along with Hewlett-Packard, another Palo Alto firm, Fairchild Semiconductor was the seed of what would become known as Silicon Valley.

From the mid-1960s into the early '70s, Fairchild Semiconductor Corporation and Texas Instruments Inc. were the leading manufacturers of integrated circuits (ICs) and were continually increasing the number of electronic components embedded in a single silicon wafer, or chip. As the number of components escalated into the thousands, these chips began to be referred to as integration chips.

Intel named their integrated chip the 4004, which referred to the number of features and transistors it had. These included memory, input, output, control, arithmetic, and logical capacities. It came to be called a microprocessor. It is this chip that is referred to as the brain of the personal, desktop computer -- the central processing unit, or CPU. Over 100,000 calculators were powered by the 4004 chip. Intel gained exclusive rights to the 4004 design, and began marketing the chip to other manufacturers in 1971.

Two young programmers realized almost immediately the need for more effective operational control of computers. Childhood friends Bill Gates and Paul Allen were whiz kids as they grew up in Seattle, Washington, debugging and improving operational control of computers at the ages of 13 and 15. As teenagers they started a company, built the hardware, and wrote the software that would provide statistics on traffic flow from a rubber tube strung across a highway. Allen quit his job and Gates left Harvard University, where he was a student, in order to create a version of the programming language "Basic" that could run on a new computer. They licensed their version and started calling their partnership Microsoft. The Microsoft Corporation went on to develop versions of Basic for nearly every computer that was released. It also developed other high-level languages. When IBM decided to enter the microcomputer business in 1980, it called on Microsoft for both a programming language and an operating system, and the small partnership was on its way to becoming the largest software company in the world.

The similar story of Apple Computer is a key part of Silicon Valley folklore. Two whiz kids, Stephen Wozniak and Steve Jobs, shared an interest in electronics. Wozniak was an early participant at Homebrew Computer Club meetings, which Jobs also occasionally attended.

Wozniak purchased one of the early microprocessors and used it to design a computer. When Hewlett-Packard, where he had an internship, declined to build his design, he shared his project at a club meeting, where Jobs suggested that they could sell it together. Their initial plans were modest. Jobs figured that they could sell it for $50, twice what the parts cost them, and they could sell hundreds of them to hobbyists. The product was actually only a printed circuit board. It lacked a case, a keyboard, and a power supply. Jobs got an order for 50 of the machines from the owner of one of the industry's first computer retail stores. To raise the capital to buy the parts they needed, Jobs sold his minibus and Wozniak his calculator. They met their 30-day deadline and continued production in Job's parents' garage.

After their initial success, Jobs sought out the kind of help that other industry pioneers had shunned. While he and Wozniak began work on the Apple II, he consulted with a venture capitalist and enlisted an advertising company to aid him in marketing. In 1976, a retired semiconductor company executive helped write a business plan for Apple, lined up credit from a bank, and hired a serious businessman to run the venture. Apple was clearly taking a different path from its competitors. For instance, while other microcomputer start-ups ran advertisements in technical journals, Apple ran an early color ad in *Playboy* magazine. Its executive team lined up nationwide distributors. Apple made sure each of its subsequent products featured an elegant, consumer-style design. This distinctive approach resonated well in the marketplace.

In 1980 the Apple III was introduced. For this new computer Apple designed a new operating system. The flagship Apple II and successors made Apple into the leading personal computer company in the world. In 1980 it announced its first public stock offering, and its young founders became instant millionaires. After three years in business, Apple's revenues had increased from $7.8 million to $117.9 million.

COMPUTER SOFTWARE

Software is a collection of computer programs and related data that provide the instructions for telling a computer what to do and how to do it. The term was coined to contrast to the old term "hardware" (meaning physical devices).

IBM knew that software was needed to make a computer useful. It contracted with several software companies to develop applications: a word processor, a spreadsheet program, and a series of business programs. Personal computers were just starting to gain acceptance in businesses, and in this market IBM had a built-in advantage, as expressed in its adage: "*Nobody was ever fired for buying from IBM.* It named its product the personal computer, which quickly was shortened to the PC. It was an immediate success, selling more than 500,000 units in its first two years.

The Software industry has grown from a few visionaries operating out of their garage with prototypes; Steve Jobs and Bill Gates were the Henry Ford and Louis Chevrolet of their times. Software falls in three categories: system software, programming software, and application software. System software provides the basic functions for computer usage and helps run the computer hardware and system. Programming software provides tools to assist a programmer in writing computer programs. Application software performs any task that benefits from computation.

One can look at the development of the computer as occurring in waves. The first large wave was the mainframe era, when many people had to share single machines. The age of mainframes could be characterized by the expression *Many persons, one computer*.

The second wave was brought on by the personal computer, which in turn was made possible by the invention of the microprocessor. The second wave can be described as the age of *One person, one computer*. The third wave may be characterized as *One person, many computers*. Since the introduction of the first personal computer, the semiconductor business has grown into a $120 billion worldwide industry.

COMPUTER HARDWARE, AND CHIP MANUFACTURE

Before integrated circuits were invented, computers used circuits of individual transistors and other electrical components -- resistors, capacitors, and diodes -- soldered to a circuit board. In 1959 Texas Instruments and Fairchild filed patents for integrated circuits. Texas Instruments found how to make all the circuit components out of germanium, the semiconductor material then commonly used for transistors. Fairchild used silicon, which is now almost universal, and found a way to build the interconnecting wires as well as the components on a single silicon chip, thus eliminating all soldered connections except for those joining the integrated circuit to other components.

The integrated circuit is formed on a silicon wafer cut from a cylinder of pure silicon 8 to 12 inches in diameter. A photographic image is made, and photolithography is used to etch successive layers as the transistor is formed in a multi-phase process. Silicon dioxide (glass) is used as electrical insulation between layers. When the fabrication is complete, a final layer of insulating glass is added, and the wafer is sawed into individual chips. Each chip is tested, and those that pass are mounted in a protective package with external contacts. [77]

The size of transistor elements continually decreases in order to pack more on a chip. As sizes decreased, electron beam or X-ray techniques became necessary. Each advance requires new fabrication plants, costing several billion dollars apiece. [78]

The principle computer elements are integrated circuits -- small silicon wafers or chips -- that contain thousands of transistors that function as electrical switches. In 1965 Gordon Moore, one of the founders of Intel, stated what has become known as Moore's Law: *The number of transistors on a chip doubles about every 18 months.* Moore's Law has been remarkably accurate for over a decade

Applied Materials Inc. is the largest supplier of manufacturing equipment, services and automation software to the semiconductor industry. Their products can be found in every semiconductor factory in the world, enabling the production of every advanced microchip. As the industry moves from one technology generation to the next in its drive to make smaller, faster and more functional chips, Applied Materials provides the leading-edge technology to help make it happen. They are the worldwide dominate supplier of equipment the chip manufacturers utilize to manufacture their chips.

COMPUTER MEMORY

The punch card was the primary means of memory during the early days of the computer. An operator would feed a stack of punched cards into the computer that would utilize this means of memory. It was never a satisfactory means, but all that existed. Magnetic core memory became the computer memory of choice throughout the 1960s, until it was replaced by semiconductor memory. A computer's memory can be viewed as a list of cells into which numbers can be placed or read. Each cell has a numbered "address" and can store a single number. The information stored in memory may represent practically anything: letters, numbers, even computer instructions.

Computer memory comes in two principal varieties: random-access memory (or RAM), and read-only memory (or ROM). ROM is typically used to store the computer's initial start-up instructions. In general, the contents of RAM are erased when the power to the computer is turned off, but ROM retains its data indefinitely.

THE INTERNET

A communication device that connects the computer with the outside world is the common telephone modem (from modulator demodulator). Modems transform a computer's digital message into an analog signal for transmission over standard telephone networks,

Computers have been used to coordinate information between multiple locations since the 1950s. The U.S. military's SAGE system was the first large-scale example of such a system, and in the 1970s, computer engineers throughout the United States began to link their computers together. In time, the network spread beyond military institutions and became known as the Internet. Initially these facilities were available to people working in high-tech environments, but in the 1990s the spread of applications like e-mail and the World Wide Web (www) saw networking become almost ubiquitous

The Internet grew out of funding by the U.S. Government to develop a communication system among government and academic computer-research laboratories. The first network component became operational in 1969. All these various networks were able to communicate with one another because of two shared protocols: the Transmission-Control Protocol, and the Internet Protocol, an addressing system that controlled the routing of packets.

I was an early user of the internet in 1980 when I was Corporate Manager of Quality Assurance for OI. An issue arose involving a possible carcinogen we used as a solvent with our plastishield label. This prompted me to seek the assistance of a PhD in Biology who was on my Quality Assurance staff. He was the only person in the company who knew how to enter and use the internet at that time. Utilizing the internet, he found the technical information we needed. The substance was not a carcinogen, but I decided in order to avoid future PR problems, we should begin immediately to phase out its use.

Originally created as a closed network for researchers, the Internet was suddenly a new public medium for information. It became the home of virtual shopping malls, bookstores, stock brokers, newspapers, and entertainment. Schools were "getting connected" to the Internet, and children were learning to do research in novel ways. The combination of the Internet, e-mail, and small and affordable computing and communication devices began to change many aspects of society.

MY COMPUTER DEVELOPMENT

I was able to utilize this new computer technology during the 1960's in the factory in Oakland where I was a manager, and I became a leader of the effort to install computerization wherever possible throughout Owens-Illinois. When I moved to corporate headquarters in Toledo, I continued to lead the computerization program throughout our factories in the United States and world-wide. The Owens-Illinois computerization of the packaging industry became one of the major success stories of the digital revolution. Since we produced one-half of all containers manufactured worldwide and had affiliates in 33 countries, our impact in the mass-production industries was considerable.

I will reveal a personal secret. While I was a key player in the conception and leadership development of computers within Owens Illinois, I never did actually place my fingers on the keyboard of a computer during my working years. I pushed developments, and dealt in concepts, relying on hands-on experts for their technical expertise. It was not until after I retired that I purchased a PC computer to do my finances that I had my first "hands-on" computer experience.

COMPUTER GAMES

Computer games are nearly as old as digital computers and have steadily developed in sophistication. Chinook, a checkers program, is widely believed to be better than any human player, and the IBM Deep Blue chess program beat world champion Garry Kasparov in 1996. These programs have demonstrated the power of modern computers.

THE "CLOUD"

The newest development for computers is the "cloud". Cloud computing is the delivery of computing as a service rather than a product, whereby shared resources, software, and information are provided to computers and other devices as a utility over a network (typically the Internet).

If a company invested millions of dollars in a huge computer, it would still be limited in the capacity of its computing and memory; however, if hundreds of computers could be linked together in an array, then the capability of that new array (cloud) would become almost limitless. Some major companies are organizing this sort of "cloud computing" capability and selling it as a new service.

Cloud computing provides computation, software applications, data access, data management and storage resources without requiring cloud users to know the location and other details of the computing infrastructure. End users access cloud based applications through a web browser or a desktop while the business software and data are stored on servers at a remote location. Cloud providers strive to give better performance as if the software programs were installed locally on end-user computers. [79]

SUMMARY:

Calculators that first appeared several centuries ago as early as the renaissance, have been replaced by the modern digital computer utilized for scientific, business, and social life. Nearly everyone world-wide from the age of eight to eighty has their hands on the keyboard of a computer almost every day.

PHYSIOLOGY, THE HUMAN BODY

"Science" is defined as the possession of knowledge through study or practice. In our own personal scheme of things, what can be more important to each of us than the science of physiology – that of our own body?

Let me start with a preface discussing what we know and what we do not know about this science. A complete understanding of physiology would involve minute nuclear particles beyond our comprehension, so the underlying basis of human life remains partially a mystery. The human body is a complex organism; therefore, an attempt to review its functions must of necessity deal with over-simplification. However, that should not dissuade us from outlining the basic physiology involved.

Life starts with the genome. This double helix is formed by molecules of DNA, which is a nucleic acid containing the genetic instructions used in the development of all living organisms. The four nucleobases (molecules) found in DNA are adenine, cytosine, guanine, and thymine. Each of these is a rather simple molecule composed of hydrogen, nitrogen, oxygen, carbon, and Phosphorus. So the basis of life is founded on rather simple biochemistry. Every cell in our body contains this DNA material.

We can see a human cell under a microscope. However, the atoms that go together to make cells are an order of magnitude too small to be seen, and we can study them only indirectly by other scientific methods. Sub-microscopic atoms are made from electrons, protons, and neutrons, which are made of particles that are yet another order of magnitude smaller -- quarks, mesons, bosons, and a host of others. High energy atom smashers of various kinds are required to detect these. At the very bottom of this menagerie of particles, field, or "whatever" that supports human life, we do not know what we may find -- or perhaps never find.

The thrust of this is that while we have learned empirically a substantial knowledge about what goes together to support our physical body -- our physiology -- we do not know much about the "why" or "how" at the sub-microscopic level. [80] As we proceed with a look at physiology it is important for us to remember what we really know, and also what we do not know.

<u>Human Cell</u>: Following that sobering preamble, we will begin our discussion of the human body with a look at the human cell. Why the cell? Because it is what every part of the body is composed of; cells make up the sum-total of the body. Cells are what constitute every organ and tissue.

"Human beings begin with the union of a male and female sex cell, each containing half the genetic material that will shape the new being. Cells that form by divisions of this initial fertilized cell orient themselves and assume different shapes and functions. Through this process, the original cell evolves by a pre-coded plan into a complex, closely integrated community of some 100 trillion cells that make up the adult human body. Each cell contains the genetic identity of the individual, yet the functions of cells forming different organs and tissues are restricted to those needed for the whole organism" [81]

Why do we breathe? It is to provide oxygen to every cell in the body. Why does the body require good blood circulation? It is so a proper level of oxygen and nutrients will reach every individual cell, and also to remove the used-up waste from the cell. Good physiology must begin with good care of the cell. Why do we engage in exercise? It is to increase the circulation of blood throughout the body so that oxygen and nutrients will reach every cell in the body; hence, maintain better health and fitness.

<u>Human Body</u>: is the entire structure of a human organism, and consists of a head, neck, torso, two arms, two legs and all the organs. By the time the human reaches adulthood, the body consists of close to 100 trillion cells, the

basic unit of life. These cells are organized biologically to eventually form the whole body. Systems include the following: cardiovascular, respiratory, immune, endocrine, lymphatic, digestive, urinary, nervous system, musculoskeletal, reproductive, and integumentary system (skin). [82]

CARDIOVASCULAR SYSTEM

This comprises the **heart** and the vascular system. The function of the heart is to circulate the **blood** that carries oxygen and nutrients to the cells throughout the body. The left side of the heart pumps blood to all parts of the body, while the right side pumps only to the lungs for re-oxygenation of the blood.

The vascular system is made up of the vessels that carry blood and lymph through the body. The arteries and veins deliver oxygen and nutrients to the body tissues and take away tissue waste matter.

The lymph vessels carry lymphatic fluid (a clear, colorless fluid containing water and a few blood cells that originates in many organs and tissues). The lymphatic system helps to protect and maintain the fluid environment of the body by filtering and draining lymph away from each region of the body. [83]

Blood moves through the circulatory system as a result of being pumped out by the heart. Blood leaving the heart through the arteries is saturated with oxygen. The arteries divide into smaller and smaller branches. As blood moves through the capillaries, the oxygen and other nutrients move out into the cells, and waste matter from the cells moves into the capillaries. As the blood leaves the capillaries, it moves into the veins, which become larger to carry the blood back to the heart.

In addition to circulating blood and lymph throughout the body, the vascular system also functions as an important component of other body systems, including: respiratory system, digestive system, kidneys and urinary system, and temperature control. In the respiratory system, blood flows through the lungs where oxygen is picked up and carbon dioxide is expelled. In the digestive system, food is digested and blood flows through the intestinal capillaries to pick up nutrients such as glucose, vitamins, and minerals. In the kidney and urinary system, waste materials from the body tissues are filtered out and leave the body in the form of urine. Regulation of the body's temperature is assisted by the flow of blood among the different parts of the body, and heat is produced by the process of breaking down nutrients for energy.

Blood pressure is the **pressure** exerted by circulating **blood** upon the walls of **blood vessels**. It is one of the body's principle **vital signs**. During each heartbeat, it varies between a maximum and a minimum pressure. It is measured on the inside of an **elbow** at the **artery**, which is the upper arm's major blood vessel that carries blood away from the heart. A person's blood pressure is usually expressed in terms of the systolic pressure over diastolic pressure.

Classification of blood pressure for adults: [84]

	Systolic	Diastolic
Desirable:	90-119	60-79

A reading above 140/90 mmHg is considered hypertension. While average values vary for any given population, there is often a large variation from person to person. It also varies in individuals from moment to moment, and varies with exercise, emotional reactions, sleep, digestion and time of day -- it is typically higher in the morning than later in the day. Also, various factors such as age and gender influence average values. In children, the normal ranges are lower than for adults and depend on height. As adults age, systolic pressure tends to rise and diastolic tends to fall. In the elderly, blood pressure tends to be above the normal adult range, largely because of reduced flexibility (elasticity) of the arteries. It is considered too low only if noticeable symptoms are present, such as light headedness.

What is vascular disease? It is a condition that affects the arteries and veins. Most often, vascular disease affects blood flow, either by blocking or weakening blood vessels, or by damaging the valves that are found in veins. Organs and other body structures may be damaged by vascular disease as a result of decreased or blocked blood flow. Causes of vascular disease include atherosclerosis, embolus, thrombus, inflammation, and trauma/injury. Atherosclerosis is a build-up of plaque, cholesterol, cellular waste products, and calcium in the inner lining of an artery. Embolus is a tiny mass of debris or a thrombus (blood clot) that may block a blood vessel. Inflammation of blood vessels includes a range of disorders and may lead to narrowing of blood vessels. Trauma or injury may lead to inflammation or infection, which can damage the blood vessels and lead to narrowing or blockage. [85]

RESPITORARY SYSTEM

The respiratory system supplies oxygen to the body. It includes airways, lungs, and the respiratory muscles. Oxygen and carbon dioxide are exchanged, by diffusion, between the atmosphere and the blood in the lungs. Red blood cells are the principal means of delivering oxygen to the body tissues via the blood flow through the circulatory system. They take up oxygen in the lungs. The lungs are the respiration organ, and the two lungs are located near the backbone on either side of the heart. The exchange of gases is accomplished in the mosaic of specialized cells that form millions of tiny, exceptionally thin-walled air sacs. Once air progresses through the mouth or nose, it travels through the larynx, the trachea, and a progressively subdividing system of bronchi until it finally reaches the lungs where the gas exchange of carbon dioxide and oxygen takes place. The oxygen is picked up by red blood cells and carried in the blood throughout the body. The intake and expulsion of air is driven by muscular action in the lungs. Medical terms related to the lung often begin with pulmo, such as in pulmonary. [86]

IMMUNE SYSTEM

The immune system, which is made up of special cells, proteins, tissues, and organs, is the body's defense against infections and other invaders. Through a series of steps called the immune response, it attacks substances that invade body systems and cause disease. The cells involved are white blood cells (leukocytes) that seek out and destroy disease-causing organisms or substances.

Leukocytes are produced or stored in many locations in the body, including the thymus, spleen, and bone marrow. There are also clumps of lymphoid tissue throughout the body, primarily as lymph nodes. The leukocytes circulate through the body between the organs and nodes via lymphatic vessels and blood vessels. In this way, the immune system works in a coordinated manner to monitor the body for germs or substances that might cause problems.

The two basic types of leukocytes are:

- Phagocytes are cells that chew up invading organisms.
- Lymphocytes are cells that help the body destroy them.

The most common type of phagocyte cell is the neutrophil, which primarily fights bacteria. If doctors are worried about a bacterial infection, they might order a blood test to see if a patient has an increased number of neutrophil triggered by the infection. [87]

An autoimmune disease is one in which the body's immune system attacks one's own body parts instead of fighting disease. In Rheumatoid Arthritis it attacks joints, in Lupus various organs, and in Dermatomyositis it attacks the muscles and skin. Treatment often is with an anti-suppressant drug to counteract a misbehaving immune system.

These complex issues involving the immune system seem academic, until someone becomes involved with an auto-immune disease such as my sister (rheumatoid arthritis), a grandson (leukemia), and me (Dermatomyositis); and then these terms become part of everyday conversation. Until inflicted with my disease, I knew virtually nothing about the immune system; now the immunologist has become my lifeline.

LYMPHATIC SYSTEM

The lymphatic system is a part of the immune system that carries a clear fluid called lymph. The lymphatic organs play an important part in the immune system, particularly the lymph nodes, and in the lymphoid follicles associated with the digestive system such as the tonsils. The system also includes the spleen, thymus, bone marrow and the lymphoid tissue associated with the digestive system. [88]

The *spleen* is an organ with important roles in regard to red blood cells and the immune system. It is located in the left upper quadrant of the abdomen. It removes old red blood cells and holds a reserve of blood in case of hemorrhagic shock while also recycling iron.

Bone marrow is the flexible tissue found in the interior of bones and produces new blood cells. On average, bone marrow constitutes 4% of the total body mass of humans. It produces approximately 500 billion blood cells per day, which utilize a conduit through bones to the body's systemic circulation. Bone marrow is also a key component of the lymphatic system, producing the lymphocytes that support the body's immune system.

ENDOCINE SYSTEM

Although we rarely think about them, the glands of the endocrine system and the hormones they release influence almost every cell, organ, and function of our bodies. The endocrine system is instrumental in regulating mood, growth and development, tissue function, and metabolism, as well as sexual function and reproductive processes.

In general, the endocrine system is in charge of body processes that happen slowly, such as cell growth. Faster processes like breathing and body movement are controlled by the nervous system. But even though the nervous system and endocrine system are separate systems, they often work together to help the body function properly.

The foundations of the endocrine system are the hormones and glands. As the body's chemical messengers, hormones transfer information and instructions from one set of cells to another. Although many different hormones circulate throughout the bloodstream, each one affects only the cells that are genetically programmed to receive and respond to its message. Hormone levels can be influenced by factors such as stress, infection, and changes in the balance of fluid and minerals in blood.

The _thyroid_ is one of the largest endocrine glands. It is found in the neck, below the thyroid cartilage (which forms the Adam's apple). The thyroid gland controls how quickly the body uses energy, makes proteins, and controls how sensitive the body is to other hormones. It participates in these processes by producing thyroid hormones that regulate the rate of metabolism and affect the growth and rate of function of many other systems in the body. Output from the thyroid is regulated by a hormone produced by the anterior pituitary, which itself is regulated by a hormone produced by the hypothalamus. The most common problems of the thyroid gland consist of an overactive thyroid gland, referred to as hyperthyroidism, and an underactive thyroid gland, referred to as hypothyroidism. [89]

A _gland_ is a group of cells that produces and secretes chemicals. It selects and removes materials from the blood, processes them, and secretes the finished chemical product for use somewhere in the body. Some types of glands release their secretions in specific areas. For instance, exocrine glands, such as the sweat and salivary glands, release secretions in the skin or inside of the mouth. Endocrine glands, on the other hand, release more than twenty major hormones directly into the bloodstream where they can be transported to cells in other parts of the body. [90]

Diseases of the endocrine system are common, including conditions such as diabetes mellitus, thyroid disease, and obesity.

DIGESTIVE SYSTEM

Digestion is the system used in the human body for the digestion of liquids and foods. It consists of the digestive tract, which are the organs through which food and liquids pass during their processing into nutrients that are absorbed into the bloodstream. It also consists of organs that contribute juices necessary for the digestive process, and structures involved in the elimination of body waste. The digestive tract begins at the lips and ends at the anus. It consists of the mouth, teeth, tongue, esophagus, stomach, small intestine, large intestine, colon, pancreas, liver and the gallbladder. [91]

Food enters the mouth, is chewed, and mixed with saliva, which contains 90% water and 10% made up of many enzymes from the salivary glands. Then the food travels down the esophagus into the stomach. The stomach is a small pouch with walls made of thick, elastic muscles, which stores and helps break down food. The enzymes in the stomach work at a specific pH and temperature. After some time (typically an hour or two), the resulting thick liquid will go into the small intestine, where 95% of absorption of nutrients occurs, and it is mixed with three different liquids:

- Pancreatic juices made by the pancreas.
- Intestinal enzymes
- Bile produced by the liver and then stored in the gallbladder.

Blood containing the absorbed nutrients is carried away from the small intestine and goes to the liver for filtering, removal of toxins, and nutrient processing.

After the food has passed through the small intestine, the food enters the large intestine. Within it, digestion is retained long enough to allow fermentation due to the action of gut bacteria, which breaks down some of the substances that remain after processing in the small intestine. The large intestine is roughly 1.5 meters long, with three parts: the junction with the small intestine, the colon, and the rectum. [92]

The _pancreas_ is a gland organ in the digestive and endocrine system. It is a long, narrow gland that is situated transversely across the upper abdomen, behind the stomach and the spleen. It is both an endocrine gland producing several important hormones, including insulin and glucagon, and also a digestive organ, secreting pancreatic juice containing digestive enzymes that assist the absorption of nutrients and the digestion in the small intestine. These enzymes help to further break down the carbohydrates, proteins, and lipids. [93]

Diabetes is a disease in which a person has high blood sugar, either because the pancreas does not produce enough insulin, or because cells do not respond to the insulin that is produced. This high blood sugar produces the classical symptoms of frequent urination, increased thirst, and increased hunger. There are three types of diabetes:

- Type 1 diabetes results from the body's failure to produce insulin and requires the person to inject insulin.
- Type 2 diabetes results from insulin resistance, a condition in which cells fail to use insulin properly, sometimes combined with an insulin deficiency.
- Gestational diabetes is when pregnant women have a high blood glucose level during pregnancy.

All forms of diabetes have been treatable since insulin became available, and type 2 diabetes may be controlled with medications. Both type 1 and 2 are chronic conditions that usually cannot be cured. Diabetes without proper treatments can cause complications such as cardiovascular disease and retinal damage. Adequate treatment of diabetes is thus important, as well as blood pressure control and maintaining a healthy body weight.

The *liver* is not only the largest gland in the body but also the most complex in function. The major functions of the liver are to participate in the metabolism of protein, carbohydrates, and fat; to synthesize cholesterol and bile acids; to initiate the formation of bile; to engage in the transport of bilirubin; to metabolize and transport certain drugs; and to control transport and storage of carbohydrates.

The body is continuously exposed to damage by viruses, bacteria, and parasites; ingested toxins and chemicals, including drugs and food additives; and foreign protein of plant origin. These insults are received by the skin, the respiratory system, and the digestive system (particularly the liver), which constitute the interface between the sterile body interior and the environment. A doctor may indicate the need for a blood test in conjunction with a physical examination, and one reason for this is to determine if toxicity is causing any damage to the liver.

URINARY SYSEM

The urinary system produces, stores, and eliminates urine. It includes two kidneys, the bladder and the urethra. The _kidneys_ are bean-shaped organs that lie in the abdomen, just below the ribcage and close to the spine. The two organs are about the size of a human fist. Situated on each kidney is an adrenal gland. The kidneys receive their blood supply from the renal arteries which are fed by the abdominal aorta. This is important because the kidneys' main role is to filter water soluble waste products from the blood. The other attachment of the two kidneys is at their functional endpoints the ureters, which run down to the urinary bladder. The kidneys perform a number of tasks, such as concentrating urine, regulating electrolytes, and maintaining acid-base homeostasis. [94]

NERVOUS SYSTEM:

How the nervous system works is exceedingly complex and there is much we do not know. At the beginning of this chapter I discussed the fact that the basis of life involves sub-microscopic particles that are orders of magnitude too small to be seen or even visualized by scientists; so, we do not know exactly how things work. That is particularly true with the nervous system.

The nervous system contains a network of specialized cells called neurons that coordinate our actions and transmit signals between different parts of the body. It consists of two parts, central and peripheral. The central nervous system contains the brain, spinal cord, and retina. The peripheral nervous system consists of sensory neurons, clusters of neurons called ganglia, and nerves connecting them to each other and to the central nervous system. These regions are all interconnected by means of neural pathways.

Neurons send signals to other cells as electrochemical waves travelling along thin fibers called axons, which cause chemicals called neurotransmitters to be released at junctions called synapses. A cell that receives a synaptic signal may be excited, inhibited, or otherwise modulated. Sensory neurons are activated by physical stimuli impinging on them, and send signals that inform the central nervous system of the state of the body and the external environment. Motor neurons situated either in the central nervous system or in peripheral ganglia, connect the nervous system to muscles or other organs. Central neurons make all of their input and output connections with other neurons. The interactions of these form circuits that generate an organism's perception of the world and determine its behavior. [95]

THE BRAIN

The brain is the most complex organ in the body and the center of the nervous system. It is the biggest user of energy of any organ in the body. In a typical human the cerebral cortex (the largest part) is estimated to contain 15 to 33 billion neurons, each connected by synapses to several thousand other neurons. These neurons communicate with one another by means of long fibers called axons, which carry trains of signal pulses to distant parts of the brain or body targeting specific recipient cells.

From a philosophical point of view, what makes the brain special in comparison to other organs is that it forms the physical structure that generates the mind. As Hippocrates put it: *"The brain gives us joys, delights, laughter and sports; and sorrows, grief's, despondency, and lamentations."*

The operations of individual brain cells are now understood to a degree, but the way they cooperate has been very difficult to decipher. The most promising approaches treat the brain as a biological computer, very different in mechanism from electronic computers, but similar in the sense that it acquires information from the surrounding world, stores it, and processes it in a variety of ways.

How do we remember the way to a friend's house? Why do our eyes blink without us ever thinking about it? Where do dreams come from? Our brain is in charge of these things and a lot more. In fact, it is the boss of our body. It runs the show and controls just about everything we do, even when we're asleep. The brain has many different parts that work together, such as these, which are key players on the brain team;

- Cerebrum
- cerebellum
- brain stem
- pituitary gland
- hypothalamus

The <u>cerebrum</u> is the biggest part of the brain that makes up 85% of the brain's weight, and is the big mass of gray matter on the periphery. It is the thinking part of the brain and controls voluntary muscles -- the ones that move when we want them to. When thinking hard, you're using your cerebrum. You need it to solve math problems and draw a picture. The memory lives in the cerebrum -- both short-term memory (what you ate for dinner last night) and long-term memory (the name of that summer trip you took ten years ago). The cerebrum also helps you reason.

The cerebrum has two halves, with one on either side of the head. The right half helps you think about abstract things like music, colors, and shapes, while the left half is said to be more analytical, helping you with math, logic, and speech. Scientists do know for sure that the right half of the cerebrum controls the left side of your body, and the left half controls the right side.

The _cerebellum_ is at the back of the brain, below the cerebrum. It's a lot smaller than the cerebrum at only 1/8 of its size. It controls balance, movement, and coordination. Because of your cerebellum, we can stand upright, keep our balance, and move around.

The _brain stem_ sits beneath the cerebrum and in front of the cerebellum. It connects the rest of the brain to the spinal cord. The brain stem is in charge of all the functions your body needs to stay alive, like breathing air, digesting food, and circulating blood. Part of the brain stem's job is to control the involuntary muscles -- the ones that work automatically, without even thinking about it. There are involuntary muscles in the heart and stomach, and it's the brain stem that tells your heart to pump more blood when you're biking or your stomach to start digesting your lunch. The brain stem also sorts through the millions of messages that the brain and the rest of the body send back and forth.

The _pituitary gland_ is very small -- only about the size of a pea! Its job is to produce and release hormones into your body. If your clothes from last year are too small, it's because your pituitary gland released special hormones that made you grow. This gland is a big player in puberty too. This is the time when boys' and girls' bodies go through major changes as they slowly become men and women, all thanks to hormones released by the pituitary gland. This gland also plays a role with lots of other hormones, and it helps keep your metabolism going. Your metabolism is everything that goes on in your body to keep it alive and growing and supplied with energy, like breathing, digesting food, and moving your blood around.

The _hypothalamus_ is like your brain's inner thermostat. The hypothalamus knows what temperature your body should be (about 98.6° Fahrenheit or 37° Celsius). If your body is too hot, the hypothalamus tells it to sweat. If you're too cold, the hypothalamus gets you shivering. Both shivering and sweating are attempts to get your body's temperature back where it needs to be.

The _amygdala_ in the brain runs the emotions. Where do feelings come from: your brain, of course? Your brain has a little bunch of cells on each side called the amygdala. The word is Latin for almond, and that's what this area looks like. Scientists believe that the amygdala is responsible for emotion. [96]

VERTEBRAL COLUMN

The backbone or spine is a column that houses and protects the spinal cord in its spinal canal. There are normally thirty-three vertebrae in humans, including the five that are fused to form the sacrum (the others are separated by discs) and the four coccygeal bones that form the _tailbone_.

MUSCULO-SKELETAL SYSTEM

A musculoskeletal system (also known as the locomotor system) is what gives humans the ability to move using the muscular and skeletal systems. This system provides form, support, stability, and movement to the body. It is made up of the body's bones (skeleton), muscles, cartilage, tendons, ligaments, joints, and other connective tissue that supports and binds tissues and organs together. The system's primary functions include supporting the body, allowing motion, and protecting vital organs.

REPRODUCTION SYSTEM

The reproductive system is the organs that work together for the purpose of reproduction. Many fluids, hormones, and pheromones are important accessories to the reproductive system. Unlike most organ systems, the two sexes often have significant differences. These allow for a combination of genetic material between two individuals, which allows for the possibility of greater genetic fitness of the offspring. The major organs of the reproductive system include the external genitalia (penis and vulva) as well as a number of internal organs including the testicles and ovaries.

INTEGUMENTARY (SKIN) SYSTEM

The skin is the organ along with hair and nails that protects the body from damage. It has a variety of functions; it serves to waterproof, cushion, and protect the deeper tissues, excrete wastes, regulate temperature, and is the site to detect pain, sensation, pressure, and temperature. The integumentary system also provides for vitamin D synthesis.

The skin is the largest of the body's organs. It accounts for about 12 to 15 percent of total body weight. It is composed of three layers of tissue: epidermis, dermis, and hypodermis. The epidermis forms the outermost layer, providing the initial barrier to the external environment. The dermis contains connective tissues, vessels, glands, follicles, hair roots, sensory nerve endings, and muscular tissue. The hypodermis is the deepest layer, which is primarily made up of adipose tissue.

The skin has an important job of protecting the body and acts as the body's first line of defense against infection, temperature change, and other challenges to stability. Functions include: [97]

- Protect the body's internal living tissues and organs
- Protect against invasion by infectious organisms
- Protect the body from dehydration
- Protect the body against abrupt changes in temperature
- Help excrete waste materials through perspiration
- Act as a receptor for touch, pressure, pain, heat, and cold
- Protect the body against sunburns by secreting melanin
- Generate vitamin D through exposure to ultraviolet light
- Store water, fat, glucose, and vitamin D
- Maintenance of the body form
- Formation of new cells to repair minor injuries

PHYSIOLOGY SUMMARY

The discipline of physiology views the body as a collection of interacting systems, each with its own combination of functions and purposes. Each body system contributes to the stability of other systems and of the entire organism. No organ or system of the body works in isolation, and the well-being of a person depends upon the well-being of all the interacting body systems. Proper nutrition and a healthy fitness program are vital for maintaining a person's physiology. What science can be more important to each of us than the health of our own physiology? Amen!

NUTRITION

Nutrition is the science most neglected by the general population. Stand on a street corner or entrance to a public school and count the percentage of adults and children who are over-weight or obese. The numbers are staggering. Something is obviously wrong with our nation's approach to nutrition. Two-thirds of adults and a third of children are overweight or obese, government statistics show. [98] On this morning's news broadcast it was reported that in another 15 years nearly 40% of the nation will be obese and many of these will be candidates for diabetes.

Why? Because we are provided with the wrong information about nutrition; hence, there is a credibility gap – plus "no one is going to tell me what to eat." The U.S. Department of Commerce with their Food Guide Pyramid is one of the principle culprits – both in what foods they recommend and what they fail to warn against. We are eating too much of the wrong foods, and also consuming way too much.

We love sugar and sweet foods, so we tend to ignore any recommendation to reduce consuming them; we have become a nation of sugar addicts. The sugar lobby is one of the most powerful special interest groups in congress. Look at how much sugar is in a bottle of cola consumed every day by children. Stroll down the supermarket aisle and read the amount of sugar contained in most breakfast cereals. Is it any wonder we are addicts who need a daily "sugar fix".

It isn't that scientist don't know better based on research; here is a sample of their findings:

> "So what should we eat? The latest clinical trials suggest that all of us would benefit from fewer (if any) sugars and fewer refined grains (bread, pasta) and starchy vegetables (potatoes). This was the conventional wisdom through the mid-1960's and then we turned the grains and starches into heart-healthy diet foods and the USDA enshrined them in the base of it famous Food Guide Pyramid as the staples of our diet. That this shift coincided with the obesity epidemic is probably not a coincidence. As for those of us who are overweight, experimental trials - the gold standard of medical evidence - suggest that diets that are severely restricted in fattening carbohydrates and rich in animal products - meat, eggs, cheese and green leafy vegetables are arguably the best approach, if not the healthiest diet to eat. Not only does weight go down when people eat this, but heart disease and diabetes risk factors are reduced. While ethical arguments against meat-eating are valid for some; health arguments against it can no longer be defended." [99]

The USDA has even established the cause of the obesity epidemic. David Wallinga of the Institute for Agriculture and Trade Policy tells us the following:

> "The USDA has established the cause of the epidemic and its 'an increase in our calorie consumption over the last 30, 35 years,' he also tells us where those calories come from: a quarter come from added sugars, a quarter from added fats (most of which are from soy), and almost half is from refined grains, mainly corn starches, wheat, and the like. ... Consumption of meat is not the problem. The same USDA data clearly shows that red-meat consumption peaked in this country in the mid-1970, before the obesity epidemic started. ... It's been dropping ever since." [100]

With that sobering introduction, let me start this chapter again and address the science of Nutrition. It is utilizing food to sustain human life, and is a science because it requires knowledge to accomplish it successfully.

Good nutrition enhances our "quality of life". Many common health problems can be prevented or alleviated with a proper diet. A poor diet can have an injurious impact on health, causing deficiency diseases such as scurvy, and such health-threatening conditions as obesity, cardiovascular disease, and diabetes. These medical conditions are not just theoretical: I can vouch for several of them. I enjoy reasonably good health and am in my mid-eighties; however, I also take prescription medications to deal with several medical conditions such as hypertension, gout, type two diabetes, and a serious auto-immune disease.

Let me relate a nutrition problem (scurvy) I encountered when I was a graduate student at U.C. Berkeley. I was living alone in a garage apartment where I cooked meals on a hot plate. My diet consisted of hamburgers, and my college lifestyle was "burning the candle at both ends". I developed a problem of mouth odor, and no amount of time with a tooth brush or mouth wash solved it. Since I was spending Tuesday nights at naval reserve drills on Treasure Island (to earn enough money to stay in college), I went to the naval dentist to seek a solution to my bad breath. He looked into my mouth and asked me questions about my life style and diet. Then he gave me his diagnosis: I had border-line scurvy as a result of poor diet. He prescribed vitamin C plus a multi-vitamin, and to start eating a healthy diet with some citrus fruit. Within a few days, my bad breath had disappeared. That was sixty years ago, and to this day I still take vitamin C and a multi-vitamin every day. It was not for another forty years that the American Medical Association recommended the use of vitamins.

As a science, nutrition has provided few champions; it simply developed slowly without direction over many centuries. Humans evolved as omnivorous hunter-gatherers during the past 250,000 years. Their diet varied depending on location and climate. The diet in the tropics tended to be based on plant foods, while at higher latitudes tended more towards animal products. Agriculture developed about 10,000 years ago in multiple locations throughout the world, providing grains, potatoes, and maize with staples such as bread. Farming provided dairy products and increased the availability of meats and the diversity of vegetables. [101]

Around 400 BC, Hippocrates said, *"Let food be your medicine and medicine be your food."* In the 16th century, Leonardo da Vinci compared metabolism to a burning candle. In 1747, Dr. James Lind, a physician in the British navy, performed the first scientific nutrition experiment, discovering that lime juice saved sailors at sea from scurvy, a deadly and painful bleeding disorder; British sailors became known as "limeys." The essential vitamin C contained in lime juice would not be identified by scientists until the 1930s. Around 1770, Antoine Lavoisier, the "Father of Nutrition and Chemistry" discovered the details of metabolism, demonstrating that the oxidation of food is the source of body heat. In the early 19th century, the elements carbon, nitrogen, hydrogen and oxygen were recognized as the primary components of food, and methods to measure their proportions were developed.

In 1912, Casimir Funk coined the term vitamin, a vital factor in the diet, from the words "vital" and "amine" because these unknown substances preventing scurvy, beriberi, and pellagra, were thought to be derived from ammonia. In 1913, Elmer McCollum discovered the first vitamins: vitamin A, vitamin B, and named vitamin C as the then-unknown substance preventing scurvy. In 1927, Adolf Otto Reinhold Windaus synthesized vitamin D, for which he won the Nobel Prize in Chemistry in 1928. In 1935 Albert Szent-Györgyi synthesized vitamin C and won a Nobel Prize.

There are four principle nutrients: carbohydrates, fats, proteins, and water. These provide the body with structural material and energy. They are measured in calories. Carbohydrates and proteins provide 4000 calories of energy per gram, and fats provide 9000 calories per gram. Another class of dietary material, fiber (ie. non-digestible material) is required for mechanical reasons to help move food through the intestines.

Carbohydrates constitute a large part of foods such as vegetables, fruit, bread, and other grain-based products, and supply much of the energy for the body.

Dietary fat consists of fatty acids bonded to a glycerol. [102] They are typically found as triglycerides. Fats may be classified as saturated or unsaturated. Unsaturated fats may be further classified as monounsaturated or polyunsaturated. Trans fats are a type of unsaturated fat that are created in an industrial process called hydrogenation, are detrimental to health, and should be avoided. Saturated fats (typically from animal sources) have been a staple in many world cultures for millennia. Unsaturated fats (e. g., vegetable oil) are healthier.

Proteins are the basis of many body structures (e.g. muscles, skin, and hair). They also form the enzymes that control chemical reactions throughout the body. The body requires amino acids [103] to produce new proteins and replace damaged proteins. Sources of dietary protein include meats, soy-products, eggs, legumes, and dairy products such as milk and cheese.

Dietary fiber is a carbohydrate that is incompletely absorbed. Whole grains, fruits (especially plums, prunes, and figs), and vegetables are good sources of dietary fiber. There are many health benefits of a high-fiber diet. It helps reduce problems such as constipation and diarrhea by increasing the weight and size of stool and softening it. Insoluble fiber, found in whole wheat flour, nuts and vegetables, helps stimulate digesta along the digestive tract. Soluble fiber, found in oats, peas, beans, and many fruits, slows the movement of food through the intestines, which may help lower blood glucose levels because it slows the absorption of sugar. Additionally, fiber is thought to possibly help lessen insulin spikes, and therefore reduce the risk of type 2 diabetes.

In addition to the principle nutrients, the body also needs minerals and vitamins. Dietary minerals are chemical elements that are present in nearly all organic molecules. Many are essential in small quantity and play a role as electrolytes. In alphabetical order they are:

- *Calcium*, needed for muscle and digestive system health, and bone strength.
- *Chlorine* an electrolyte.
- *Magnesium*, builds bone, increases flexibility.
- *Phosphorus*, component of bones; essential for energy.
- *Potassium*, an electrolyte (heart and nerve health).
- *Sodium*, excessive sodium consumption can deplete calcium and magnesium, leading to high blood pressure.
- *Sulfur*, in many proteins (skin, hair, nails, liver, and pancreas).
- *Vitamins*: Some vitamins are recognized as essential nutrients, necessary for good health. Thousands of phyto-chemicals [104] have recently been discovered in food (particularly in fresh vegetables). Vitamin deficiencies may result in disease conditions, including scurvy, osteoporosis, impaired immune system, certain forms of cancer, symptoms of premature aging, and poor psychological health.

- *Water*: is excreted from the body in multiple forms; including urine and feces, sweating, and by exhaled breath; therefore, it is necessary to adequately rehydrate to replace lost fluids. Some recommendations for the quantity of water required for maintenance of good health suggested that 6–8 glasses of water daily is the minimum to maintain proper hydration; however, the notion that a person should consume eight glasses of water per day cannot be traced to a credible scientific source. Much of the needed quantity is contained in prepared foods. For healthful hydration, the current guidelines recommend total water intake of 2.0 liter/day. This includes drinking water, other beverages, and from food.
- *Antioxidants*: As cellular metabolism requires oxygen, potentially damaging compounds known as free radicals can form, which must be neutralized by antioxidant compounds. Some are produced by the human body and others may be obtained in the diet. Phytochemicals [105] make up the majority of antioxidants. These nutrients are typically found in edible plants, especially colorful fruits and vegetables, and blackberries are a source of polyphenol antioxidants.

The Mediterranean Diet is based on the observation that people of the Mediterranean region suffer a relatively low incidence of coronary heart disease, despite having a diet relatively rich in saturated fats; however, statistics collected more recently show that the incidence of heart disease in those countries may have been underestimated and, in fact, may be similar to that of many other countries. Diets that become popular fads should sometimes by challenged.

Mental - Cognitive issues: research indicates that improving nutrition and establishing long-term habits of healthful eating have a positive effect on cognitive memory capacity, potentially increasing a person's potential to process and retain information. Nutritional treatment may be appropriate for major depression, bipolar disorder, schizophrenia, and obsessive compulsive disorder, the four most common mental disorders in developed countries.

Insulin Resistance: Several lines of evidence indicate reduced insulin function is a decisive factor in various diseases. For example, it is linked to chronic inflammation, which in turn is linked to developments such as arterial micro-injuries, heart disease and cancer. Insulin resistance is characterized by a combination of abdominal obesity, elevated blood sugar, elevated blood pressure, elevated blood triglycerides, and reduced HDL cholesterol. Obesity clearly contributes to insulin resistance, which in turn can cause diabetes. Although the association between overweight and insulin resistance is clear, the exact causes of insulin resistance remain less clear; however, it has been demonstrated that appropriate exercise, regular food intake, and reducing glycemic foods can reverse insulin resistance in overweight individuals.

The updated USDA Food Guide Pyramid published in 2005 is a general nutrition guide for recommended food consumption. Nutritional standards are established jointly by the US Department of Agriculture and US Department of Health and Human Services. Dietary and physical activity guidelines from the USDA are presented in the concept of a food pyramid, which superseded the Four Food Groups. Unfortunately, these governmental guidelines from the US Department of Agriculture have a strong political input, reflecting the inputs of the various interest groups in agriculture such as the citrus and grain industries, and its credibility has been challenged. There is the indication that these governmental guidelines may not represent the best nutritional information.

In the US, dietitians are registered with the Commission for Dietetic Registration and the American Dietetic Association, and are only able to use the title "dietitian," as described by the business and professions codes of each respective state. In California, registered dietitians must abide by the "Business and Professions Code of Section 2585-2586.8"

A 1985 US National Research Council report entitled Nutrition Education in US Medical Schools concluded that nutrition education in medical schools was inadequate. Only 20% of the schools surveyed taught nutrition as a separate, required course. Little has changed in the years since; your family physician is seldom a strong information resource for nutrition.

As stated at the beginning of the chapter, good nutrition enhances our "quality of life." What is more enjoyable and healthy than a family Sunday dinner with a glass of red wine, antipasta salad, entre of veal scaloppini, fresh vegetables, dessert of cheese and blueberries, and accompanied with great conversation. That is nutrition at its best.

PHYSICAL FITNESS

It may be a stretch to call physical fitness a science, but no more so than calling nutrition a science because of its relation to physiology, or calling mental health a science because it is an important component of psychology. In the case of physical fitness, it is a factor that has been proven to extend the life span of people in addition to adding to their quality of life. Science is defined: "Something (as a sport or technique) that may be studied or learned like a systematized knowledge." [106] Yes, physical fitness is a science.

Fitness needs to be started early in a child's life as a normal part of growing up. I was fortunate to be raised in a small, rural town where we walked everywhere and played numerous games (including during school recess in sunshine and snow). My siblings and I spent many Saturdays climbing through the mountains west of Buffalo Gap. We did all those things because they were fun and were supported by our parents and the local community.

It is one thing to understand the physiology of the human body, but it is something else to take care of it by getting proper exercise. Unfortunately, too many Americans fall in this latter category. Exercise does not come easy. It requires hard work and makes a person sweat; however, the tradition for exercise is contained in our genes. Our ancestors, who shaped our genes, got their exercise in normal daily activities: working in the fields, hunting in the forests, and walking everyplace. Natural exercise -- exertion -- was a way of life for them. Few people now engage in these pioneer things, leading to our modern sedentary lifestyle.

Physical exercise is any bodily activity that enhances or maintains physical fitness and overall health and wellness. It is performed for various reasons including strengthening muscles and the cardiovascular system, honing athletic skills, weight loss or maintenance, as well as for the purpose of enjoyment. Frequent and regular physical exercise boosts the immune system, and helps prevent the "diseases of affluence" such as heart disease, cardiovascular disease, Type 2 diabetes and obesity. It also improves mental health, helps prevent depression, helps to promote or maintain positive self-esteem, and can even augment an individual's sex appeal or body image, which is also found to be linked with higher levels of self-esteem, Childhood obesity is a growing global concern and physical exercise may help decrease some of the effects of childhood and adult obesity. Health care providers often call exercise the "miracle" or "wonder" drug -- eluding to the wide variety of proven benefits that it provides.

Physical exercises are generally grouped into three types, depending on the overall effect they have on the human body: [107]

- Flexibility *exercises*, such as stretching, or improving the range of motion of muscles and joints.
- Aerobic exercises, such as cycling, swimming, skipping rope, running, playing tennis, or cardiovascular endurance.
- Anaerobic exercises, such as weight training, functional training, and eccentric training or sprinting.

My daily program (5 days per week) at the local fitness club includes each of their 3 types of exercise.

Physical exercise is important for maintaining physical fitness and can contribute positively to maintaining a healthy weight, building and maintaining healthy bone density, muscle strength, and joint mobility, promoting physiological well-being, reducing surgical risks, and strengthening the immune system.

Frequent and regular aerobic exercise has been shown to help prevent or treat serious conditions such as high blood pressure, obesity, heart disease, Type 2 diabetes, insomnia, and depression. According to the World Health Organization, lack of physical activity contributes to approximately 17% of heart disease and diabetes, 12% of falls in the elderly, and 10% of breast cancer and colon cancer. [108]

Exercise reduces levels of cortisol, [109] (a steroid hormone) that causes many health problems, both physical and mental.

There is evidence that vigorous exercise is more beneficial than moderate exercise. Some studies have shown that vigorous exercise executed by healthy individuals can increase opioid peptides [110] that are responsible for exercise-induced euphoria, increase testosterone and growth hormone, effects that are not as fully realized with moderate exercise. Both aerobic and anaerobic exercise work to increase the mechanical efficiency of the heart by increasing cardiac volume. Such a change is healthy if it occurs in response to exercise.

I am personally a strong believer in vigorous exercise and participated in an aerobic exercise group into my mid-eighties. I now have modified the program because I must take an immune suppressant medication that makes me vulnerable to infectious diseases of group activities, so I attempt to do similar exercises at home.

Not everyone benefits equally from exercise. There is variation in individual response to training; where most people will see a moderate increase in endurance from aerobic exercise, some individuals will as much as double their oxygen uptake, while others can never augment endurance. The genetic variation in improvement from training is one of the key physiological differences between elite athletes and the larger population. Studies have shown that exercising in middle age leads to better physical ability later in life. Tennis is an excellent exercise at any age.

The beneficial effect of exercise on the cardiovascular system is well documented. There is a direct relation between physical inactivity and cardiovascular mortality, and inactivity is a risk factor for the coronary artery disease. The greatest potential for reduced mortality is in the sedentary who become moderately active. Most beneficial effects of physical activity on cardiovascular disease mortality can be attained through moderate-intensity activity. ... Persons who modify their behavior after a heart attack to include regular exercise have improved rates of survival. ... Persons who remain sedentary have the highest risk for all-causes of disease mortality. [111]

Although there have been hundreds of studies on exercise and the immune system, there is little direct evidence on its connection to illness, but some suggests that moderate exercise has a beneficial effect on the immune system, and moderate exercise has been associated with a decreased incidence of upper respiratory tract infections. Inflammations which are associated with chronic diseases are reduced in active individuals, and the positive effects of exercise may be due to its anti-inflammatory effects. [112]

A 2008 review of cognitive enrichment therapies (strategies to slow or reverse cognitive decline, ie. Alzheimer's) concluded that physical activity enhances older adults' cognitive function. In addition, physical activity has been shown to reduce the risk of developing dementia. Furthermore, anecdotal evidence suggests that frequent exercise may reverse alcohol-induced brain damage. [113]

There are several possibilities for why exercise is beneficial for the brain. Examples are:

- Increasing the blood and oxygen flow to the brain
- Increasing growth factors that help create new nerve cells.
- Increasing chemicals in the brain that help cognition, such as dopamine and serotonin

Physical activity is thought to have other beneficial effects related to cognition as it increases levels of nerve growth factors, which support the survival and growth of a number of neuronal cells.

A number of factors may contribute to depression including being overweight, low self-esteem, stress, and anxiety. Endorphins act as a natural pain reliever and antidepressant in the body. Endorphins [114] have long been regarded as responsible for what is known as "runner's high", a euphoric feeling a person receives from intense physical exertion. When a person exercises, levels of both circulating serotonin and endorphins are increased. These levels are known to stay elevated even several days after exercise is discontinued, possibly contributing to improvement in mood, increased self-esteem, and weight management. Exercise is a potential prevention method for treatment of mild forms of depression. Research has shown that when exercise is done in the presence of other people (familiar or not), it can be more effective in reducing stress than simply exercising alone. [115]

A 2010 review of published scientific research suggested that exercise generally improves sleep for most people, and helps sleep disorders such as insomnia. The optimum time to exercise may be 4 to 8 hours before bedtime, though exercise at any time of day is beneficial, with the possible exception of heavy exercise taken shortly before bedtime, which may disturb sleep. There is, in any case, insufficient evidence to draw detailed conclusions about the relationship between exercise and sleep.

Too much exercise can be harmful. Without proper rest, the chance of stroke or other circulation problems increases, and muscle tissue may develop slowly. Extremely intense, long-term cardiovascular exercise, as can be seen in athletes who train for multiple marathons, has been associated with scarring of the heart and heart rhythm abnormalities. Inappropriate exercise can do more harm than good, with the definition of "inappropriate" varying according to the individual. For many activities, especially running and cycling, there are significant injuries that occur with poorly regimented exercise schedules. Injuries from accidents also remain a major concern. [116]

Stopping excessive exercise suddenly can also create a change in mood. Feelings of depression can occur when withdrawal from the natural endorphins produced by exercise occurs. Exercise should be controlled by each body's inherent limitations. While one set of joints and muscles may have the tolerance to withstand marathons, another body may be damaged by 20 minutes of light jogging. This must be determined for each individual. Too much exercise can also cause a female to miss her period, a symptom known as amenorrhea. [117]

Proper nutrition is as important to health as exercise. When exercising, it becomes even more important to have a good diet to ensure that the body has the correct nutrients to aid the body with the recovery process following strenuous exercise.

The benefits of exercise have been known since antiquity. Marcus Cicero, around 65 BC, stated: "*It is exercise alone that supports the spirits, and keeps the mind in vigor.*" However, the link between physical health and exercise (or lack of it) was only discovered in 1949 and reported in 1953 by a team led by Jerry Morris. Dr. Morris noted that men of similar social class and occupation (bus conductors versus bus drivers) had markedly different rates of heart attacks, depending on the level of exercise they got: bus drivers had a sedentary occupation and a higher incidence of heart disease, while bus conductors were forced to move continually and had a lower incidence of heart disease. This link had not previously been noted and was later confirmed by other researchers. [118]

ECONOMICS
Management of a Household

Two years ago I published a book on economics that was well received in many quarters, *RATIONAL MARKET ECONOMICS: A Compass for the Beginning Investor.* [119] That focus on investing was not scientific in keeping with the theme of this present book, because finance is not a "science". So, let me now place economics within a broader context. According to the definition in my tattered college dictionary: "Economics is a social science concerned with analysis of the production, distribution, and consumption of goods and services." [120]

Economics is a Greek word relating to "management of a household," and it starts in the gut of the individual with how people react to the events in their lives. This notion did not exist several centuries ago before people had "connected the dots," and realized how things in their lives -- needs and resources --were interrelated.

An Englishman, Adam Smith, became the first economist when he published *An Inquiry into the Nature and Causes of the Wealth of Nations* [121] in 1776, the same year that Thomas Jefferson wrote our Declaration of Independence. No one had realized until then how the fragments of social activity fit together in a cohesive whole. Smith accomplished that and the result was a blueprint for a new science called economics, which analyzes the products, distribution, and consumption of goods and services. It explains how things interact not only in business, finance, and government; but also in such social aspects as education, law, crime, politics, and war. In his book, Adam Smith opened up the entire range of social discourse.

Economics is generally dealt with on two levels: *macroeconomics* and *microeconomics*. Macroeconomics looks at the big picture and deals with performance, decision-making and social structure. It analyzes the entire economy and issues affecting it, including unemployment, inflation, economic growth, monetary policy, and fiscal policy. Microeconomics comes from the Greek word meaning "small," and focuses on the details. While we ordinarily think of this only within the narrow context of finances, it is important to realize that the entire range of society is impacted and intertwined with finances, such things as the standard of living and crime. The reason we look at both macro and micro economics is because both have considerable impact on how our society works.

There are two opposing philosophies of how much control should be maintained over the financial aspects of the economy: *Free Market*, and *Keynesian*. Free Market relies on minimal economic intervention and regulation by the state, except to enforce private contracts and the ownership of property. Keynesian advocates monetary policies by the central bank and fiscal policies by the government to stabilize the business cycle. [122] While these two opposing philosophies may seem somewhat academic, we realized during the 2007 domestic and global financial meltdown their differences can create great impact in the social environment.

Two important parameters of economics are *supply* and *demand*. These two factors determine how prices vary as a result of a balance between product availability and demand in the economy. Supply is the relation between the price of a good and the quantity available for sale. Demand is the relation of the quantity that all buyers would be prepared to purchase at that price. Quantity and price are normally inversely related; that is, the higher the price, the less of it people would be prepared to buy. Other factors can change demand; for example, an increase in income may increase the demand. Market equilibrium occurs where quantity supplied equals quantity demanded.

The supply and demand relationship includes the distribution of income among capital and labor; hence, determines the "standard of living" in a society. In the labor market the quantity of labor employed and the wage rate depends on the supply and demand for labor. Economics examines the interaction of workers and employers to explain patterns of wages, mobility, unemployment, productivity, and related public-policy issues. Supply-and-demand analysis is used to explain the behavior of competitive markets and macroeconomic variables such as Gross Domestic Product (GDP), which is one of the yardsticks used to measure the economy. [123]

Gross Domestic Product is the market value of all goods and services produced within a country in a given period of time. All countries like to maintain their GDP in a positive direction. Recent United States administrations have a goal to maintain it stabilized within the 2% to 4% annual growth rate. When the GDP growth rate is below zero for two consecutive quarters, an economy is considered to be in recession. It seldom climbs above the 5% annual growth rate in the United States, but the GDP of a number of emerging foreign countries such as China often climb up to a 10% annual growth rate because they have started from a lower base.

Economies are seldom entirely stable but fluctuate over several months or years, involving shifts between periods of rapid growth (expansion or boom) and periods of stagnation or decline (contraction or recession). These are referred to as the business cycle. Business cycles are usually measured by considering the GDP growth rate. Despite being termed cycles, these fluctuations in economic activity do not follow a predictable periodic pattern.

An important aspect of economics is the financial picture and particularly as it relates to investment and the stock market. The economy is a complex social phenomenon with many dimensions, and some of these measure important parameters of the stock market. The most important are these:

- Corporate Earnings
- Interest Rates
- Inflation
- Liquidity
- $ Exchange rate

An investor who keeps focus on these economic parameters will be in a better position to predict the future course of the market, or at least be able to react more intelligently to events as they arise. There are also many secondary factors that affect the market, but they do this mostly through the influence that they exert on these primary parameters. Here is an overview:

CORPORATE EARNINGS: Investors buy stock to gain part ownership of a corporation so they can share in its earnings, or in growth that will lead to future earnings. The price they are willing to pay for stock is based on current earnings and their perception of future earnings compared to what they could make from other investments.

INTEREST RATES: Interest rates and the stock market have a tendency to move in opposite directions because of two reasons:

- (One) Interest bearing investments are competitors of equity stocks for available investor's dollars.
- (Two) Interest rates are used by the Federal Reserve Bank as a tool to fuel or to retard the economic growth rate, and this has a major impact on corporate earnings.

The net effect is that the stock market normally reacts to changes in interest rates faster and more sharply than most other factors. Several interest rates are important to economic markets: prime rate set by commercial banks, discount rate and federal funds rate set by the Federal Reserve Bank, and Treasury bills determined by bids in the open market.

INFLATION: The Federal Reserve Bank considers inflation one of the greatest threats to economy stability, so they place high priority on its control. One of the tools they use is interest rates. As inflation increases, interest rates are raised by the Federal Reserve Bank as a means to cause inflation to be decreased.

LIQUIDITY: Liquidity means how much money supply and other liquid assets exist that are readily available to the economy. It involves both how much is available and how fluid it is to flow into various investments. When liquidity is too low, the economy and the stock market suffer. When liquidity is high and investor's pockets are full of money, they tend to buy stock; so the markets often rise.

<u>U.S.DOLLAR EXCHANGE RATE</u>: Until recent years it made little difference to the investor how much a U. S. dollar was worth compared to the Euro or other foreign currencies. That has changed in recent decades because we now deal with a global economy.

These five primary economic parameters are interrelated. As one of them goes up or down it may cause another parameter to move in parallel or in the opposite direction. Seldom are all five parameters stable or in lock step.

Economics, unlike some other sciences, is based primarily on subjective evaluations using economic data. The controlled experiments common to some physical sciences are difficult and uncommon in economics, and instead broad data is observationally studied; this type of testing is less rigorous than controlled experimentation, and the conclusions more tentative. Statistical methods such as regression analysis are common. Practitioners use such methods to estimate the size, economic significance, and statistical significance of inter-relations. By such means, a hypothesis may gain acceptance, although in a probabilistic, rather than certain sense. The methods need not produce a final conclusion or even a consensus on a particular question, given different tests, data sets, and prior beliefs.

Subjective results inherent in economics have led some to argue that economics is not a "genuine science." Since mathematics, physics, biology, and virtually all sciences occasionally involve subjective or only probabilistic results, the debate is inconsequential. I will pass judgment on the controversy: *economics is a science.*

WEATHER & CLIMATE

In South Dakota we encountered extremes of weather: 110° F in the summer, -40° F in the winter, heavy snow, searing droughts, and occasional floods. However, it was all normal for us, and other than an occasional topic of conversation, weather was never of great concern. Everyone knew what would be, would be; and whatever came along, we had been there before and survived it nicely -- despite any discomfort. There were things to worry about during the Great Depression of the 1930's other than something over which we had no control.

When I was a midshipman during pre-flight training at a naval training base in Pensacola, Florida, I attended classes designed to instruct us in weather as it may pertain to aviation. The classes were held at 6:30am to avoid mid-day heat, and it was a struggle to stay awake, so I'm not sure how much I actually learned. The films had been made for the military during World War Two and represented state-of- art knowledge about weather systems at that time, and we successfully fought World War Two using that quality of weather information. The most famous forecasting was for the invasion at Normandy by Allied forces. An unusually intense June storm brought high seas and gales to the French coast, but a moderation of the weather that was successfully predicted enabled General Dwight D. Eisenhower, supreme commander of the Allied Expeditionary Forces, to make his critical decision to invade on D Day, June 6, 1944.

Weather is the state of the atmosphere. It involves such phenomena as temperature, humidity, precipitation, air pressure, wind, and cloud cover. Weather differs from climate in that the latter is the synthesis of weather conditions that prevail during a long time period -- generally 30 years.

Weather occurs in the troposphere, the lowest region of the atmosphere that extends from the Earth's surface up to 5 miles at the poles and to about 11 miles at the Equator. Although weather is largely confined to the troposphere, jet streams and air of higher regions significantly affect sea-level patterns. Geographic features, most notably mountains and large bodies of water, also affect weather. Research has revealed that ocean-surface temperatures are a potential cause of atmospheric temperature anomalies in successive seasons and at distant locations. One such manifestation of interactions between the ocean and the atmosphere is the El Niño. It is responsible not only for unusual weather events in the Pacific region (Australia and western South America) but also for those that occur in Europe and the United States. The El Niño phenomenon appears to influence mid-latitude weather conditions by modulating the position and intensity of the polar-front jet stream.

The changeability of weather varies widely in different parts of the world. It is most pronounced in the mid-latitudes, where a continuous procession of high- and low-pressure centers produces a constantly shifting weather pattern. In tropical regions, by contrast, weather varies little from day-to-day or from month-to-month.

Weather has a tremendous influence on human settlement patterns, food production, and personal comfort. Extremes of temperature and humidity cause discomfort and may lead to the transmission of disease; heavy rain can cause flooding, displacing people and interrupting economic activities; thunderstorms, tornadoes, and hail storms may damage or destroy crops and buildings. In coastal areas, hurricanes and typhoons can cause great damage. The absence of rainfall can cause droughts and dust storms when winds blow over parched farmland, as with the "dustbowl" conditions of the U.S. Plains states in the 1930s. [124]

The Greek philosophers had much to say about meteorology. Unfortunately, they probably made many bad forecasts, because Aristotle, who was the most influential, did not believe that wind is air in motion. He did believe, however, that west winds are cold because they blow from the sunset.

The early scientific study of meteorology is associated with the invention of the mercury barometer by Torricelli, an Italian physicist in the mid-17th century, and the development of a reliable thermometer. A succession of achievements by chemists and physicists of the 17th and 18th centuries contributed meteorological research: the gas laws by Boyle, and the law of partial pressures by Dalton were scientific breakthroughs that made it possible to understand aspects of the atmosphere and its behavior. During the 19th century, all of these earlier ideas began to produce results in terms of useful weather forecasts.

Weather forecasting has become a science in recent decades. It is the prediction of the weather through application of the principles of physics, supplemented by a variety of statistical and empirical techniques. However, despite the scientific means, observations remain vital in the case of weather forecasting. From the days when early humans ventured from caves, individuals survived by being able to detect nature's signs of impending snow, rain, or wind, indeed of any change in weather. With such information they enjoyed greater success in the search for food and safety, the major objectives of that time. During my youth in South Dakota, the same was also true for farming. Farmers with the ability to predict the weather in advance of the season were usually the most successful in farming. The *Farmer's Almanac* with its weather information was in constant use by successful farmers. [125]

In a sense, weather forecasting is still carried out in basically the same way as it was by the earliest humans -- namely, by making observations and predicting changes. The tools used to measure temperature, pressure, wind, and humidity in the 21st century obviously are better; yet, even the most sophisticated forecast made on a supercomputer requires a set of measurements of the condition of the atmosphere -- an initial picture comparable to that formed by our forebears when they looked out of their cave dwellings. The primeval approach entailed insights based on the experience of the observer, while the modern technique consists of the additional step of solving equations in a computer. Although seemingly quite different, there are underlying similarities between both. [126]

The *synoptic* weather map came to be the principal tool of 19th-century and continues to be used today in weather reports around the world. (Synoptic is derived from the Greek word "comprehensive".) It characterizes the weather and prevailing conditions over a large region.

Since the mid-20th century, digital computers have made it possible to calculate changes in atmospheric conditions so that scientists can obtain the same result from the same initial conditions. Technological advances since the 1960s have led to a growing reliance with Earth-orbiting satellites. By the late 1980s, forecasts of weather were largely based on numerical models integrated by high-speed supercomputers. [127]

A major breakthrough in measurement came with the launching of the first meteorological satellite by the United States in 1960. The impact of global views of temperature, cloud, and moisture distributions, as well as of surface properties (*e.g.,* ice cover and soil moisture), has already been substantial. New methods are making the 21st century the "age of the satellite" in weather prediction.

Scientists frequently advance ideas before the technology exists to implement them. Few better examples exist than numerical weather forecasting. These forecasts are objective calculations of changes to the weather map based on sets of equations called models. Human forecasters may interpret the results of the computer models, but there are few forecasts that do not begin with numerical-model calculations of pressure, temperature, wind, and humidity.

The computer model begins with an algorithm, which is a step-by-step procedure for calculations. Starting from an initial state, the instructions describe a computation that will proceed through a number of well-defined successive states. The equations in the forecasting algorithm are a set of differential equations that are used to approximate global atmospheric flow. [128] They consist of three main sets of equations:

- *Conservation of momentum*: Equations that describe flow on the surface of a sphere
- *Thermal energy*: Relating the temperature to heat sources
- *Continuity equation*: Representing the conservation of mass

In general, these equations relate the five variables u, v, ω, T, W and their evolution over time and space

u: The velocity in the east/west direction.
v: The velocity in the north/south direction.
Ω: The vertical isobaric velocity
T: The temperature.
W: The perceptible water.

The basic idea of numerical weather prediction is to use the equations of fluid dynamics and thermodynamics at a given time to estimate the state of the fluid at some future time. The main inputs are surface observations from automated weather land stations and buoys at sea. The World Meteorological Organization acts to standardize the instrumentation, observing practices and timing of these observations worldwide. Sites launch radiosondes, which rise through the depth of the troposphere and well into the stratosphere. Meteorological radar provides information on precipitation location and intensity. Additionally, if weather radar is used then wind speed and direction can be determined. [129]

Long-range weather forecasting has a different approach from the computerized forecasting just discussed. In most cases, long-range forecasters use the climatologically approach and concern themselves with the weather picture over a longer period of time. The limit of day-to-day forecasts based on the "initial map" approach is about two weeks. The atmosphere is a chaotic system, so small changes to one part of the system can grow to have large effects on the system as a whole. This makes it difficult to accurately predict weather more than a few days in advance. Most long-range forecasts thus attempt to predict the departures from normal for a given month or season. The National Weather Service expresses predictions in probabilistic terms, making it clear that they are subject to uncertainty. Forecasts of temperature are more reliable than those of precipitation, monthly forecasts are better than seasonal ones, and winter months are more accurate than other seasons.

Since the mid-1980s with the advent of supercomputers, interest has grown in applying computer models to long-range forecasting -- beyond 20 or 30 days. The reliability of long-range forecasts has improved substantially in recent years; yet, many significant problems remain unsolved, posing challenges for those engaged in the field of weather forecasting.

Moving from a discussion of weather to climate, the forecast methods of the latter are less scientific and non-computerized. Climate change is the variation in global or regional climates over time; reflecting changes over time scales ranging from decades to millions of years. These can be caused by processes internal to the Earth, external forces such as sunlight intensity, or human activities. In recent usage, the term "climate change" refers to changes in modern climate, including the rise in average surface temperature known as global warming.

The Earth has undergone periodic climate shifts in the past, including four major ice ages. These consisted of glacial periods where conditions were colder than normal, separated by interglacial periods. Snow and ice during a glacial period reflects more of the Sun's energy into space, cooling the Earth. Increases in greenhouse gases can increase the global temperature and produce an interglacial period. Suggested causes of ice age periods include the positions of the continents, variations in the Earth's orbit, changes in the solar output, and volcanism.

Climate models use quantitative methods to simulate the interactions of the atmosphere, oceans, land surface and ice, but remain in their infancy with questions as to their accuracy. They are used for a variety of purposes from study of the dynamics of the weather and climate system to projections of future climate. All models attempt to balance incoming energy of electromagnetic radiation with outgoing energy of long-wave electromagnetic radiation from the Earth. Any imbalance theoretically results in a change in the average temperature of the Earth. The most talked-about applications of these models in recent years have been their use to infer the consequences of greenhouse gases in the atmosphere, primarily carbon dioxide. These models predict an upward trend in the global mean surface temperature, with the most rapid increase in temperature being projected for the higher latitudes of the Northern Hemisphere.

Perhaps someday in the future, science will make the same advances in climate forecasting that we already have in weather forecasting.

THE SCIENCE INVOLVING ENERGY

Energy has over-reaching aspects in almost every phase of modern life. As the world's population increases will there be enough energy? This leads to questions such as what forms of energy will prevail, and how will this be decided: by economic or political clout, by legal restrictions, or by war? (As an example of legal restrictions: the State of California prohibits Homeowners' associations from restricting solar devices.) [130]

Let's begin with a brief review of the science. It is axiomatic that energy cannot be created or destroyed; only converted from one form to another. It exists as either potential or kinetic energy. Potential energy is that which matter has because of its position or because of the arrangement of parts. Kinetic energy is what an object possesses due to motion. [131] An example is the internal combustion engine that converts the potential chemical energy in gasoline and oxygen into heat, which is then transformed into the propulsive kinetic energy that moves a car.

There are five forms of energy available in our modern society: fossil fuel, nuclear, solar, wind, and tidal. All transformations from one energy source to another, such as from the sun shining on a solar panel to an electric current, or from the blade of a wind turbine to an electric current, involve some loss of available energy, and no conversions are 100% efficient.

FOSSIL FUELS

There are three fossil fuels: coal, oil and natural gas. All three were formed many hundreds of millions of years ago before the time of the dinosaurs -- hence the name fossil fuels. Approximately 90% of the world's electricity is generated with fossil fuels in thermoelectric generators, which typically operate at 30–40% efficiency, losing power with heat losses to the environment. These generators are utilized where efficiency and cost of fuel are less important. [132]

It is estimated that primary sources of energy consist of petroleum 36.0%, coal 27.4%, and natural gas 23.0%, amounting to an 86.4% share for fossil fuels in primary energy consumption in the world. Non-fossil sources included hydroelectric 6.3%, nuclear 8.5%, and others (geothermal, solar, tide, wind, wood, waste) amounting to 0.9%. [133]

The burning of fossil fuels produces around 21 billion tons of carbon dioxide (CO_2) per year, but it is estimated that natural processes can only absorb about half of that amount, so there is a net increase of 10 billion tons per year. Carbon dioxide is one of the greenhouse gases that contributes to global warming, causing the average surface temperature of the Earth to rise in response, which the vast majority of climate scientists agree will cause major adverse effects.

Fossil Fuel Reserves: Coal, oil, and natural gas provide 80% of energy production: coal 35%, oil 24%, and natural gas 21%.

Proved reserves in 2007:
- Coal: 905 billion metric tons (equivalent to 4416 trillion barrels of oil.)
- Oil: 1,119 billion barrels.
- Natural gas: 6,380 trillion cubic feet (equivalent to 1,160
- Billion barrels of oil).

Years of production left in the ground with the proved reserves:
- Coal: 148 years
- Oil: 43 years
- Natural gas: 61 years

Years of production with the most optimistic reserve estimates: [134]
- Coal: 417 years
- Oil: 43 years
- Natural gas: 167 years.

As hydrocarbon supplies diminish, prices will rise. Higher prices will lead to increased alternative supplies as previously uneconomic sources become sufficiently economical to exploit. Artificial gasoline (i.e. ethanol from corn) and other renewable energy sources require more expensive production and processing technologies than conventional petroleum reserves, but may become economically viable in the near future. [135]

OIL SHALE & FRACKING

This is a recent development that holds the potential for enormous change involving the use of energy. Oil shale is a fine-grained sedimentary rock from which hydrocarbons can be produced. It originated the same as crude oil, but heat and pressure have not yet transformed it into petroleum; however, it can be processed to produce natural gas. This is done by "fracking", which is a process of hydraulic fracturing of shale to allow the trapped gas to be released. It is accomplished by pumping water, sand, and chemicals down into a well at high pressure. [136]

These oil shales exist in various places throughout the U.S. and worldwide. According to the International Energy Agency, the global use of natural gas will rise by more than 50% compared to 2010 levels and account for 25% of world energy demand by 2035. [137]

The process of fracking has environmental concerns that must be addressed since it presents the potential for contamination of ground water supplies.

INTERNAL COMBUSTION ENGINES

The internal combustion engine is an engine in which the combustion of a fuel (normally a fossil fuel) occurs with an oxidizer (usually air) in a combustion chamber. In an internal combustion engine, the expansion of the high-temperature and high-pressure gases produced by combustion apply direct force to some component of the engine. This force is applied to pistons, turbine blades, or a nozzle, and moves the vehicle, transforming chemical energy into useful mechanical energy.

The term internal combustion engine usually refers to an engine in which combustion is intermittent, such as the six-stroke piston engine. Another class uses continuous combustion, such as gas turbines, jet engines and most rocket engines. These engines are frequently powered by gasoline and are most commonly used for mobile applications and dominate for cars, aircraft, and boats.

Modern gasoline engines have a maximum thermal efficiency of about 25% to 30% when used to power a car. In other words, even when the engine is operating at maximum thermal efficiency, about 70-75% is rejected as heat without being turned into useful work, i.e. turning the crankshaft. Approximately half of this rejected heat is carried away by the exhaust gases, and half passes through the cylinder walls into the engine cooling system and is passed to the atmosphere via the coolant system radiator. Some of the work generated is also lost as friction, noise, air turbulence, and work used to turn engine equipment and appliances such as water and oil pumps and the electrical generator, and only about 25-30% of the energy released by the fuel consumed is available to move the vehicle. [138]

In the past several years, Gasoline Direct Injection increased the efficiency of the engines equipped with this fueling system up to 35%. Currently the technology is available in a wide variety of vehicles. [139]

HYBRID VEHICLES

A hybrid vehicle is a vehicle that uses two or more power sources to move the vehicle. The term most commonly refers to hybrid electric vehicles, which combine an internal combustion engine and an electric motor. The hybrid vehicle typically achieves greater fuel economy and lower emissions than conventional internal combustion engine vehicles. These savings are primarily achieved by three elements of a typical hybrid design:

(1) Relying on both the engine and the electric motors for peak power needs results in a smaller engine that can have less internal losses and lower weight.
(2) Having significant battery storage capacity to store and reuse recaptured energy, especially in stop-and-go traffic.
(3) Recapturing significant amounts of energy during braking that are normally wasted as heat.

Other techniques that are not necessarily 'hybrid' features, but that are frequently found on hybrid vehicles include:

(1) Shutting down the engine during traffic stops or while coasting or during other idle periods.
(2) Improving aerodynamics. Part of the reason that SUVs get such bad fuel economy is the drag on the car since a box shaped car has to exert more force to move through the air.
(3) Using low rolling resistance tires (tires were often made to give a quiet, smooth ride, high grip, etc., but efficiency was a lower priority). Hybrid cars use special tires that are more inflated than regular tires and so improving fuel economy.
(4) Powering the a/c, power steering, and other auxiliary pumps electrically only when needed, which reduces mechanical losses when compared with driving them continuously with traditional engine belts.

Hybrid vehicle emissions today are getting close to the recommended level set by the EPA (Environmental Protection Agency). The recommended levels they suggest for a typical passenger vehicle should be equated to 5.5 metric tons of carbon dioxide. The three most popular hybrid vehicles set even more stringent standards by producing less than 4.0 tons showing a major improvement in carbon dioxide emissions. Hybrid vehicles can reduce air emissions of smog-forming pollutants by up to 90% and cut carbon dioxide emissions in half.

SOLAR ENERGY.

My introduction to solar power was as a graduate student at U.C. Berkeley in 1956 when I worked part-time at the University Engineering Field Station in Richmond. A project located next door was a solar-to-electrical development engineered by one of the professors. Although I was not directly associated with the project, the professor shared the results with me. His conclusion: solar power was technically feasible, but not economically viable. On that basis, the project died.

Twenty years later, the viability of solar power was pursued by my company, Owens-Illinois, in the aftermath of the oil embargo by OPEC countries and the severe fuel shortage created in the United States that drove up the cost of gasoline. The company established a Solar Division that utilized its expertise in glass manufacturing to develop solar for conversion to electrical power. The project was abandoned and the division dissolved several years later as a result of the conclusion: solar power was technically feasible, but not economically viable. The fossil fuel industry had the ability to increase production whenever it suited its purpose, and drive the comparative cost with solar power out of the ballpark.

In 2010, Applied Materials began a project to develop solar power, but soon spun it off to a company named Solyndra, which went into bankruptcy when startup costs were not offset with income. While Solyndra had sufficient technical capability, it was unable to survive because it was unable to economically compete with producers from China that were receiving huge subsidies from the Chinese Government. It has become apparent several times in the past that solar power is not economically feasible unless one of three things happens: Fossil fuel becomes unavailable, fossil fuel costs increased significantly and reach parity with solar power, or solar power is supported by governmental subsidies.

Solar technical issues: Photovoltaic systems use solar panels to convert sunlight into electricity. A system is made up of one or more solar panels. A small system may provide energy to a single consumer, or to an isolated device like a lamp or a weather instrument. Large grid-connected systems can provide the energy needed by many customers. Due to the low voltage of an individual solar cell (typically 0.5v), several cells are wired in series. The electricity generated can be stored, used directly, or fed into a large electricity grid powered by central generating plants. Most use an inverter to convert the DC power produced into alternating current that can power motors. The modules are usually first connected in series to obtain the desired voltage, and the individual strings are then connected in parallel to allow the system to produce more current.

Costs of production have been reduced in recent years. Cost of solar power was about $8-10/watt in 1996, and reportedly reduced to $3-10/watt ten years later in 2006. Cost reductions have been made when crystal silicon solar cells were replaced by less expensive multi-crystalline silicon solar cells, and thin film silicon solar cells have also been developed recently at even lower costs.

Standalone systems vary widely in size and application from wristwatches or calculators to remote buildings or spacecraft. If the load is to be supplied independently, the generated power is stored and buffered with a battery. In non-portable applications where weight is not an issue, such as in buildings, lead acid batteries are most commonly used for their low cost. In small devices (e.g. calculators, parking meters) only direct current (DC) is consumed. In larger systems (e.g. buildings, remote water pumps) AC is usually required. An inverter is used to convert the DC from the modules or batteries into AC current. [140]

Solar Vehicles: Vehicles may obtain some or all of the energy required for their operation from the sun. Surface vehicles generally require higher power levels than can be sustained by a practical-sized solar array, so a battery is used to meet peak power demand, and the solar array recharges it. Space vehicles have successfully used solar photovoltaic systems for years of operation, eliminating the weight of fuel or primary batteries.

Solar Grid System: A grid connected system is connected to a large independent grid (typically the public electricity grid) and feeds power into the grid. Such systems vary in size from residential (2-10kW) to solar power stations (up to 10s of mW). This is a form of decentralized electricity generation. In the case of building mounted systems, the electricity demand of the building is met, and the excess is fed into the grid. In kW sized installations the system voltage is as high as permitted (typically 1000V). Most modules generate about 160W at 36 volts.

Photovoltaic arrays are commonly used on rooftops to supplement power use. Often the building will have a connection to the power grid, in which case the energy produced can be sold back to the utility in some sort of metering agreement. My son had a roof-mounted system in California used to heat a swimming pool, and a considerable amount of the electricity was fed into the grid, generating him a substantial savings in power costs.

A square meter photovoltaic installation in the Southwestern United States operating with an average transformation efficiency of 12% may expect to produce 1 kWh/day. In the Sahara desert, one could ideally obtain closer to 8.3 kWh/m^2/day. The unpopulated area of the Sahara desert is over 9 million km^2, which if covered with solar panels would provide 630 terawatts total power. Since the Earth's current energy consumption rate is around 13.5 TW, deserts hold great potential for the future use of solar power.

<u>Solar Insolation</u>: Insolation is a measure of solar radiation energy received on a given surface area during a given time. Accounting for clouds, the fact that most of the world is not on the equator, and that the sun sets in the evening, insolation is the correct measure of solar power -- the average number of kilowatt-hours per square meter per day. For the weather and latitudes of the United States, typical insolation ranges from 4kWh/m^2/day in northern climes to 6.5 kWh/m^2/day in the sunniest regions.

In 2010, solar panels available for consumers can have a yield of up to 19%, while commercially available panels can go as far as 27%. A photovoltaic installation may expect to produce 1 kWh/m^2/day. Solar panels have an average efficiency of 12%, with the best commercially available panels at 20%. Effective module lives are typically 25 years or more. The payback period for an investment in a PV solar installation varies greatly, but is typically calculated to be between 10 and 20 years.

<u>Solar Power Economics</u>: With the market structures and prices which have prevailed in the past, photovoltaic has not been able to compete directly in the bulk electricity market. It has relied on support mechanisms such as feed-in tariffs, loan guarantees and tax credits. However the costs have declined strongly towards grid parity, where it will be a viable source of mainstream power. [141]

WIND POWER

Wind power is not new. When I was living in South Dakota during the 1930's, nearly every rancher had a windmill in the pasture that pumped water into a tank for his cattle, and also a windmill near the ranch house. Friends of our family, the Sanson's lived in an isolated mountain region and were wealthy enough to install a wind turbine that fed into a huge battery so they had electricity 24/7 for all their home and ranch needs. We were invited to their home for an evening dinner, and I marveled at the luxury of such lighting out in the country -- the only local rancher with it. Ten years later, the REA (Rural Electrification Act) had been enacted and all the ranchers throughout the country had electricity, and most windmills vanished (or gradually fell down with deterioration).

Humans have been using wind power for at least 5,000 years to propel sailboats and sailing ships. Windmills have been used for irrigation pumping and for milling grain since the 7th century AD in what are now Afghanistan, India, Iran and Pakistan.

The total amount of available power from the wind is more than present use from all other sources. Current worldwide capacity of wind-powered generators is 238 gigawatts (GW), growing by 41 GW each year. Wind power now has the capacity to generate about 2.5% of worldwide electricity usage. Wind power market penetration is expected to reach 8 percent by 2018 in the U.S. Several countries have already achieved relatively high levels of wind power penetration, such as 21% of stationary electricity production in Denmark, 18% in Portugal, 16% in Spain, 14% in Ireland and 9% in Germany in 2010. Eighty countries around the world are now using wind power on a commercial basis.[142]

A large wind farm may consist of several hundred individual wind turbines which are connected to the electric power transmission network. Alongside the highway as I travel from my home in Sonora to San Francisco over the Altamont Pass, I can see a wind farm with 6000 wind turbines, and most afternoons they are all spinning with the high winds blowing eastward toward the hot San Leandro valley. Developed during a period of tax incentives in the 1980s, this wind farm has more turbines than any other in the United States.

Wind power, as an alternative to fossil fuels, is plentiful, renewable, widely distributed, clean, produces no greenhouse gas emissions during operation, and uses little land. The cost per unit of energy produced is similar to the cost for new coal and natural gas installations. The construction of wind farms is not universally welcomed, but any effects on the environment from wind power are generally much less problematic than those of any other power source. Large-scale implementation of wind energy production will however need to address concerns related to aesthetic and environmental factors, and land availability.

Worldwide there are now many thousands of wind turbines operating. Wind generation capacity more than quadrupled between 2000 and 2006, doubling about every three years. The United States pioneered wind farms and led the world in installed capacity in the 1980s and into the 1990s. German installed capacity surpassed the U.S. and led until once again overtaken by the U.S. in 2008. China has been rapidly expanding its wind installations and passed the U.S. in 2010 to become the world leader.[143]

In some geographic regions, peak wind speeds may not coincide with peak demand for electrical power. In the state of California, for example, hot days in summer may have low wind speed and high electrical demand due to air conditioning. An option is to interconnect widely dispersed geographic areas with a "Super grid". In the US it is estimated that to upgrade the transmission system to take in potential renewable would cost at least 60 billion.

According to a 2007 Stanford University study published in the Journal of Applied Meteorology and Climatology, interconnecting ten or more wind farms can allow an average of 33% of the total energy produced to be used, as long as minimum criteria are met for wind speed and turbine height. [144]

Wind Power Economics: Wind power has low ongoing costs, but a moderate capital cost. The cost per unit is averaged over the projected useful life of the equipment, which may be in excess of twenty years. Wind's costs have dropped recently in the range of 5 cents per kilowatt-hour, about 2 cents cheaper than coal-fired electricity. Thirty-five percent of all new power generation built in the United States since 2005 has come from wind, more than new gas and coal plants combined. [145]

Wind energy sometimes receives financial or other support to encourage its development, and subsidies in many jurisdictions. In the US, wind power receives a tax credit for each kW·h produced. Another tax benefit is accelerated depreciation. Many states also provide incentives, such as exemption from property tax, mandated purchases, and additional markets for "green credits".

Compared to the environmental impact of traditional energy sources, the environmental impact of wind power is relatively minor. Wind power consumes no fuel, and emits no air pollution. While a wind farm may cover a large area of land, many uses such as agriculture are compatible.

TIDAL POWER

Tidal power converts the energy of tides into useful forms of power – mainly electricity. Although not yet widely used, tidal power has potential for future electricity generation. Tides are more predictable than wind energy and solar power. Recent technological developments and improvements, both in design and turbine technology indicate that the total availability of tidal power may be much higher than previously assumed, and that economic and environmental costs may be brought down to competitive levels. [146]

Tidal power can be classified into two generating methods:

- Stream Generators: Tidal stream generators make use of the kinetic energy of moving water to power turbines.
- Tidal Barrage: Tidal barrages make use of the difference in height between high and low tides and are essentially dams across the full width of a tidal estuary. [147]

Whether tides will become economically viable depends on the technology of how to accomplish installations in the harsh environment of the ocean.

WATER SCIENCE

Let me tell you why I never eat Van Camp Pork and Beans: it goes back one hundred and twelve years. In 1914, my dad at the age of twelve roamed the Black Hills in the vicinity of the Cuyhoga Mine where his dad worked. One day he passed by another mine on Iron Creek owned by Van Camp, a rich guy from the east who owned the pork and bean company. His mine was temporarily idle because of a broken part in a pump. Wanting to be helpful to a neighbor, Dad told Van Camp they probably had the same part over at the Cuyhoga Mine. Van Camp urged him to go there and get it. Dad ran the 3 miles to Cuyhoga, then carried the heavy part back the three miles to the Iron Creek mine, and proudly handed it to the owner. Van Camp took it, slowly turned it over inspecting it carefully, and then said: "It looks the same". Then he flipped Dad a nickel -- <u>only a nickel</u>? And that is why no member of the Keating family has eaten a can of Van Camp Pork and Beans for one hundred and twelve years.

You may wonder why I relate this? It is because the story of water can be rather boring, and I felt the need to give the chapter some exciting front-end loading.

Maybe water isn't so boring after all. Water is an absolute necessity of life: without it humans will die. Yet, because we take it for granted, we face increasing peril. Even though we are surrounded by oceans of water, we have a shortage. The lack of adequate water may be one of the principle factors that will ultimately limit our civilization, if nuclear war does not kill us all first.

In 1800, Thomas Malthus an English scholar became widely known for his theories about population and its rapid increase by geometric proportions. His publication of An Essay on the Principle of Population observed that sooner or later population will lead to famine and disease. Malthus wrote: "*The power of population is greater than the power to produce subsistence for man. ...The increase of population is limited by the means of subsistence. ...Population increases when the means of subsistence increase, and. ...The power of population is repressed and kept equal to the means of subsistence.*" [148]

When Malthus wrote this in 1800, the population of the world had just reached one billion for the first time. It would be only another 123 years before it reached two billion in 1927, and it took only 33 years to rise by another billion people, reaching three billion in 1960. Thereafter, the global population reached four billion in 1974, five billion in 1987, six billion in 1999, and seven billion in October 2011. [149]

Water shortages are already a major factor in many areas of the world. The major cities of Los Angeles and San Diego import nearly 100% of their water through canals and pipes from as far away as 400 miles. Ground water and other local sources throughout Southern California have already been exhausted. That scenario is already repeated in major population centers throughout the world.

The term Peak Water has been put forward as a concept to help understand growing constraints on the availability, quality, and use of freshwater resources. The definitions of the term were laid out in a 2010 article in the Proceedings of National Academy of Sciences. It defines peak renewable, peak non-renewable, and peak ecological water. There is a vast amount of water on the planet but managed water is becoming scarce. Much of the world's water in underground aquifers and in lakes is being depleted; the phrase peak water sparks debates similar to those about peak oil. The concept has sparked such high interest that the New York Times chose the phrase "peak water" as one of their 33 Words of the Year for 2010.

There is concern that this condition of peak water is being approached in many areas around the world. Some areas are suffering from peak renewable water, where entire flows are being consumed for human use; peak non-renewable water, where groundwater aquifers are being over pumped faster than nature recharges them; and peak ecological water, where environmental constraints are overwhelming the economic benefits provided by water use. If present trends continue, nearly two billion people will be living with water scarcity by 2025, and two thirds of the world population could be subject to water stress

Fresh water is a renewable resource, yet the world's supply of fresh water is under increasing demand. 96.5 % percent of the world's water is salty. Nearly 70% of the fresh water is frozen in the icecaps of Antarctica and Greenland, and most of the remainder is present as soil moisture or lies in deep underground aquifers not accessible to human use. Less than 1% of the world's fresh water is accessible for direct human use: the water found in lakes, rivers, and underground sources. Only this minuscule amount is renewed by rain and snowfall, and available on a sustainable basis. [150]

Water demand already exceeds supply in many parts of the world, and more areas will experience this shortage as the world population continues to rise. The highest consumption of water comes from India, China and the United States, countries with large populations and extensive agricultural irrigation demand for food production. [151]

India has 20 percent of the Earth's population, but only four per cent of its water. Water tables are dropping fast in some of India's main agricultural areas. The Indus and Ganges rivers are tapped so heavily that, except in rare wet years, they no longer reach the sea. India has the largest water withdrawal of all the countries in the world, and eighty-six per cent of that water goes to support agriculture. [152]

China as the most populous country has the second largest water withdrawal and is facing a crisis. Sixty-eight per cent of that water goes to support agriculture, and its industrial base is consuming another twenty-six percent. One third of China's population lacks access to safe drinking water. Rivers and lakes are dying, groundwater aquifers are over-pumped, and adverse impacts on human health are widespread. [153]

The United States has 5% of the world's population, yet the U.S. uses almost as much water as India or China. Substantial amounts of water are used to grow food exported to the rest of the world. There are 36 states in the U.S. in some degree of water stress, from serious to severe.

The Ogallala Aquifer in the high plains (Texas, New Mexico, Kansas, Nebraska, and South Dakota) is being mined at a rate that exceeds replenishment -- a classic example of peak non-renewable water. Portions of the aquifer will not recharge because rainfall does not match extraction for irrigation. [154]

In California, massive amounts of groundwater are also being withdrawn from Central Valley groundwater aquifers -- unreported, unmonitored, and unregulated. California's Central Valley is home to one-sixth of all U.S. irrigated land, and the state leads the nation in agricultural production and exports, but the inability to sustain groundwater withdrawals will lead to adverse impacts on the region's agricultural productivity. [155]

The Central Arizona Project is a 336-mile long canal that diverts water from the Colorado River to irrigate more than 300,000 acres of farmland. The project also provides drinking water for Phoenix and Tucson. It has been estimated that Lake Mead, which dams the Colorado, has a 50-50 chance of running dry by 2021. [156]

The Ipswich River near Boston now runs dry in some years due to heavy pumping of groundwater for irrigation. Maryland, Virginia and the District have been fighting over the Potomac River. In drought years like 1999 or 2003 and on hot summer days, the region consumes up to 85 percent of the river's flow. [157]

WATER DESALINIZATION

Desalinization refers to processes that remove salt from saline water to produce fresh water that is suitable for human consumption or irrigation. Most of the interest in desalination is focused on developing cost effective ways of providing fresh water from the ocean for human use. This is a non-rainfall-dependent water source, but desalination uses large amounts of energy, making it more expensive than fresh water from conventional sources such as rivers or groundwater. Desalination is particularly relevant to countries that traditionally have relied on collecting rainfall behind dams to provide their drinking water supplies.

According to the International Desalination Association, 14,451 desalination plants operated worldwide in 2009, producing a year-on-year increase of 12.3% for useable water. The world's largest desalination plant is in the United Arab Emirates. The traditional process used in these operations is vacuum distillation -- essentially boiling of water at less than atmospheric pressure. A leading distillation method is multi-stage flash distillation, accounting for 85% of production worldwide. [158]

A competing process uses membranes to desalinate, applying reverse osmosis technology. These use semi-permeable membranes and pressure to separate salts from water. Reverse membrane systems typically use less energy than thermal distillation, which has led to a reduction in overall desalination costs over the past decade. Desalination remains energy intensive, however, and future costs will continue to depend on the price of both energy and desalination technology. [159]

Co-generation is the production of potable water from seawater in a dual-purpose facility in which a power plant becomes the source of energy for desalination. There are various forms of co-generation plants; however, the majority use either fossil fuels or nuclear power as their source of energy. Most plants are located in the Middle East or North Africa that use their petroleum resources to offset limited water resources.

Nuclear reactors can also be used to produce large amounts of potable water, and are in use around the world, from India to Japan and Russia. Eight nuclear reactors coupled to desalination plants are operating in Japan alone -- or were operating prior to the meltdown in 2011.

Additionally, the current trend in dual-purpose facilities is hybrid configurations, in which water from a reverse osmosis desalination component is mixed with distillate from thermal desalination; basically, two or more desalination processes are combined along with power production. Such facilities have already been implemented in Saudi Arabia at Jeddah and Yanbu.

A typical aircraft carrier in the U.S. military uses nuclear power to desalinate 400,000 US gallons of water per day. I spent three years during the Korean War on a destroyer in which our principal water supply was from desalinated water. Among other privations of shipboard life, one of the least pleasurable was the option between a 15 minute salt water shower or a 30 second fresh water shower (wet down, turn the shower off, soap and scrub, turn water back on to rinse, and don't exceed 30 seconds of fresh water use).

A 2004 study argued that desalinated water may be a solution for some water-stress regions, but not for places that are deep in the interior of a continent, or at high elevation. Unfortunately, that includes some of the places with biggest water problems. One needs to lift water up in elevation by a mile, or transport it nearly a thousand miles before transport costs are equal to desalination costs. Thus, it is often more economical to transport fresh water from somewhere else than to desalinate it. Some cities like New Delhi or Mexico City have the combination of both the cost of desalination followed by the costs of transportation.

The desalination process is spreading around the world as a solution to water shortages. Singapore is desalinating water for US$0.49 per cubic meter. The city of Perth began operating a reverse osmosis seawater desalination plant in 2006 that is powered partially by renewable energy from a wind farm. A desalination plant is now operating in Australia's largest city of Sydney. The Australian government announced that it would build a seawater desalination plant for the city of Adelaide, which will be funded by raising water rates. A utility recently won approval to build a desalination plant north of San Diego that will produce enough drinking water to supply about 100,000 homes. [160]

Improved technology has cut the cost of desalination in half in the past decade, making it more competitive. Current costs are about $950 per acre-foot, which compares with an average $700 an acre-foot that local agencies now pay for water. $1,000 per acre-foot works out to $3.06 for 1,000 gallons, or $.81 for 1 cubic meter. A utility in San Leandro, California reportedly was desalinating water for US $0.46 per cubic meter. [161]

Considerable research is underway involving water science. A Jordanian-born chemical engineering doctoral student at University of Ottawa invented a new desalination technology that is alleged to produce between 600% and 700% more water output per square meter of membrane than current technology. General Electric is looking into similar technology, and the U.S. National Science Foundation funded University of Michigan to study it as well. Patent issues and details of the technology remain unresolved. [162]

In the United States, due to a court ruling under the Clean Water Act, ocean water intakes are no longer viable without reducing fish kills (fish, plankton, and fish larvae) by 90%. So, ocean environmental concerns are also an issue. There are alternatives, including beach wells that eliminate this concern, but require more energy and higher costs while limiting output. To limit the environmental impact of returning the brine to the ocean, it can be diluted with another stream of water entering the ocean, such as the outfall of a wastewater treatment plant or power plant. Another method to reduce the increase in salinity is to mix the brine via a diffuser in a mixing zone. For example, once the pipeline containing the brine reaches the sea floor, it can split into many branches, each releasing brine gradually through small holes along its length. Mixing can be combined with power plant or waste water plant dilution. [163]

Perhaps it is time for the Keating family to end their vendetta against old-man Van Camp, since he has been deceased now for over a hundred years. I will take that under consideration; after all, a nickel is still only a nickel.

SPACE EXPLORATION

Man has visually explored outer space since earliest times and wondered what was out there with the stars, and particularly those bright objects later identified as planets. Galileo was the first to discover physical details about the individual bodies of the Solar System using a telescope he developed. He discovered that the moon was cratered, the sun was marked with sunspots, and that Jupiter had four satellites around it. [164] That was in the 17th century, and it would be another four centuries before man was able to send vehicles into space for exploration and finally go into space for a first-person experience. All the planets in our Solar System have now been visited to varying degrees by spacecraft launched from earth during this century. Man has landed on the moon and returned to earth with samples of material from there.

The first man-made objects launched into space were in 1942 by German scientists during World War II while testing the V2 rocket. After the war, the United States used German scientists and their captured rockets in programs for both military and civilian research. The Soviets, also with the help of German teams, launched sub-orbital rockets including radiation and animal experiments on some flights. The first successful launch into orbit was of a Soviet unmanned mission in 1957. In the political climate of the Cold War, the Soviet success caught the United States off-guard, which gave impetus to a major escalation of space exploration in domestic US politics. In 1958, President Eisenhower signed the National Aeronautics and Space Act, establishing NASA. It began operations by absorbing a 46-year-old predecessor organization intact; its 8,000 employees, an annual budget of US$100 million, three major research laboratories and two small test facilities. A significant contributor to the U.S. program was technology from the German rocket program led by Wernher von Braun, who had formerly led the German rocket project during the war. The first successful human spaceflight carried Russian cosmonaut Yuri Gagarin in 1961 with his spacecraft completing one orbit around the globe. Gagarin's flight resonated around the world; it was a demonstration of the advanced Soviet space program, and it opened an entirely new era in space exploration: human spaceflight. Two months later, the U.S. launched a person into space with Alan Shepard's suborbital flight, and orbital flight was finally achieved by the United States in 1962 when John Glenn orbited the Earth.

The initial driving force for development of space technology was a weapons race for intercontinental ballistic missiles (ICBMs) to be used for nuclear weapon delivery. In 1961 when USSR launched the first man into space, the U.S. declared itself to be in a "Space Race" with Russia. During the next couple decades, developments in space technology came quickly. An early pioneer in NASA was Bob Gilruth, director of 25 manned space flights, who suggested to President Kennedy that the Americans take the bold step of reaching the Moon in an attempt to reclaim space superiority from the Soviets. The first man to walk on the surface of another Solar System body was Neil Armstrong, who stepped onto the Moon in 1969; five more Moon landings occurred through 1972. The Moon program was one of the most expensive American scientific programs ever. It is estimated to have cost $202 billion in present day US$. In comparison the Manhattan (atom bomb) project cost roughly today-US$25.8 billion. [165]

The United States' Space Shuttle, which debuted in 1981, is the only spacecraft to make multiple orbital flights. The five shuttles have flown a total of 121 missions, with two of the craft destroyed in accidents. The first orbital space station was NASA's Skylab. The first human settlement in space was the Soviet space station Mir, which was continuously occupied for ten years from 1989 to 1999, and its successor, the International Space Station, has maintained a continuous human presence in space for 12 years. A Russian's record spaceflight of 438 days aboard the Mir space station has not been surpassed. Long-term stays in space reveal issues with bone and muscle loss in low gravity, immune system suppression, and radiation exposure.

After the first 20 years of exploration, focus shifted from one-off flights to renewable hardware, such as the Space Shuttle program, and from competition to cooperation as with the International Space Station.

With the substantial completion of the space station project in 2012, plans for space exploration by the USA remain in flux. A Bush Administration program for a return to the Moon by 2020 was judged inadequately funded and unrealistic by an expert review panel reporting in 2009. The US Senate and House of Representatives are still working towards a compromise NASA funding bill

While the Sun will probably not be physically explored in the near future, one of the reasons for going into space includes knowing more about the Sun. Spacecraft above the atmosphere and the Earth's magnetic field have access to the Solar wind and infrared and ultraviolet radiations that cannot reach the surface of the Earth. The Sun generates space weather that can affect power generation and transmission systems on Earth.

Other parts of our solar system have now been explored. The first successful probe to fly by another Solar System body sped past the Moon in 1959. Originally meant to impact with the Moon, it instead missed its target and became the first artificial object to orbit the Sun. Mercury remains the least explored of the inner planets and only two missions have made close observations. Due to the high energy expenditure to reach Mercury with its great velocity due to its proximity to the Sun, it is difficult to explore and orbits around it are rather unstable. Venus, despite one of the most hostile environments in the solar system, has had more "landers" sent to it than any other planet in the solar system. The exploration of Mars has been an important part of the space exploration programs, and dozens of robotic spacecraft have been launched since the 1960s, but the exploration of Mars has come at a considerable financial cost with roughly two-thirds of all spacecraft destined for Mars failing before completing their missions, a high failure rate attributed to the complexity and large number of variables involved in an interplanetary journey. The exploration of Jupiter has consisted solely of "flybys", in which detailed observations are taken without the probe landing or entering orbit. As Jupiter is believed to have only a small rocky core and no real solid surface, a landing mission is nearly impossible. Saturn,

Uranus, and Neptune have been explored only through flyby spacecraft, with no other visits currently planned. [166]

While not as spectacular, exploration of the Earth as a celestial object in its own right has been perhaps the most fruitful in providing useful knowledge of space. Orbital missions can provide data for the Earth that can be difficult or impossible to obtain from a purely ground-based point of reference. For example, the existence of the Van Allen belts was unknown until their discovery by a satellite. These belts contain radiation trapped by the Earth's magnetic fields, which currently renders construction of habitable space stations above 1000 km impractical. The hole in the ozone layer was found by a satellite that was exploring Earth's atmosphere, and satellites have allowed for the discovery of archeological sites or geological formations that were difficult or impossible to otherwise identify.

While many comets have been closely studied from Earth sometimes with centuries-worth of observations, only a few comets have been closely visited. In 1985, a satellite conducted the first comet fly-by before studying the famous Halley comet. A spacecraft smashed into another comet to learn more about its structure and composition while another mission returned samples of another comet's tail. In 2010, an unmanned spacecraft developed by the Japanese rendezvoused with a small asteroid, studied the asteroid's shape, spin, topography, color, composition, density, collected samples, and returned to the earth. [167]

Technical research that is conducted is one of the reasons supporters cite to justify government expenses of the space program. NASA programs often showed ongoing economic benefits (such as NASA spin-offs), generating many times the revenue of the cost of the program. NASA has produced a series of public service announcements supporting the concept of space exploration. Overall, the public remains largely supportive of space exploration.

Critics such as the late physicist and Nobel prize winner Richard Feynman have contended that manned space travel (as distinguished from space exploration in general, such as robotic missions) has never achieved any major scientific breakthroughs. [168]

Common rationales for exploring space include advancing scientific research, uniting different nations, ensuring the future survival of humanity and developing military and strategic advantages against other countries.

SOCIOLOGY

Sociology does not carry the prestige of some other sciences; for example, calling yourself a sociologist would be unlikely to get you seated at the head table at Sardi's. However, despite its lack of clout in scientific circles, it has become front-and-center in recent years with a commanding presence in worldwide communications and phenomena such as the Social Media, which incited the Arab Spring and is helping to restructure the Middle East.

A few years ago when an earlier book of mine was published, I was asked to sign-up for a Social Media package to assist with its promotion. The package included such unknown (to me) and esoteric names as Facebook and Twitter. Little did I realize these would soon become major worldwide phenomena? Issues of sociology have become increasingly important in our inter-connected and complex world.

The question is sometimes asked: is sociology a science? The answer is: yes, it is a science, but not a natural science. It falls under the umbrella of fields such as criminology, education, political science, and international relations. Each of these involves subjective criteria that are not always testable explanations or predictions. Nevertheless, sociology is a science that *"uses various methods of empirical investigation and critical analysis to develop a body of knowledge about human social activity."* [169]

What science is more important to us than one that addresses our inter-relationships with human behavior? The natural sciences such as chemistry, mathematics, and computers are important, but our human connections with other people get addressed by us each and every day. Sociology is difficult to characterize; by its very nature, it is based on a subjective body of knowledge that uses empirical investigation rather than the "scientific method". That does not mean its findings should not be given credence, but rather that they must be qualified and considered within the context in which they arise.

Sociological reasoning began centuries before the discipline was recognized as anything approaching a science, and included such names as Confucius, Plato, Ibn Khaldun (a 14th century Arab scholar), Auguste Comte (a French philosopher of the 18th century), and Karl Marx (German of the 19th century). In the 18th century, Adam Smith became a sociologist with his book, *An Inquiry into the Nature and Causes of the Wealth of Nations*. [170] While we think of him more in connection with economics, he connected the dots and placed it within the context of the entire social culture of his day. Sociology has now grown so large and is populated with such an international cast of practitioners that it is difficult to identify the current principal players.

Social theories lack an overarching foundation and there is little consensus about an academic framework. [171] Typical language is obscure and obtuse as to be nearly incomprehensible. Here is an example: "*This third generation of social theory includes phenomenologically inspired approaches, ethno methodology, symbolic interactionism, post-structuralism, and tradition of hermeneutics, and ordinary language philosophy.*" [172] Wow!! As a physicist am I expected to know what that means? Stick around. I will try to sort through the hyperbole and attempt to reach more understandable verbiage.

Although the subject matter in the social sciences differ from those in natural science, it is possible to organize them within areas of activity, such as the following:

<u>Social organization</u> is the study of institutions, social groups, ethnic groups, family, education, politics, and religion.

<u>Social psychology</u> is the study of human nature in relation to group life, social attitudes, collective behavior, and personality formation.

<u>Human ecology</u> deals with behavior and its relationship to social institutions,

i.e., the prevalence of mental illness, criminality, delinquencies, divorce, and drug addiction in urban centers.

Demography is the study of population and change as they influence the economic, political, and social system.

Applied sociology utilizes findings such as education, marriage, criminology, ethnic relations, family counseling, and other aspects of daily life.

Moving on from this academic approach to sociology, there are many every-day activities in which sociologist are actively engaged such as criminology, the environment, education, family life, and health. Here is a brief discussion of the current focus in each.

Criminality, deviance, Law, and Punishment. Criminologists analyze the nature, causes, and control of criminal activity, drawing upon methods of sociology, psychology, and the behavioral sciences. They focus on behaviors that violate norms: why do these norms exist; how they change over time; and how they are enforced? An example currently receiving much attention is prison incarceration. "The U.S. has 760 prisoners per 100,000 citizens, which is ten times more than most other developed countries. Japan has 63 per 100,000, Germany has 90, France 96, and Britain the highest at 153. Even developing countries that are well known for their crime problems have a third of the U.S. numbers." [173] Why do we have so many more in prisons, and what can be done about it?

Economic sociology traces its roots to Adam Smith in 1776, the same year as our Declaration of Independence. There is currently discussion of how the economic pie will be carved up within the United States among the various economic classes: the wealthy, middle class, and those in poverty.

Environmental sociology is the study of human interactions with the natural environment. Much current focus is directed at the "greenhouse effect" and its possible impact on global warming. How and by what means will society obtain the energy sources needed in the future. Other issues involve the conservation of natural resources, and maintenance of a healthy environment.

Sociology of education is particularly concerned with schooling in the United States, where students are now ranked lower than in other major countries. What is the cause and what must we do to re-establish our leadership in education? A 1966 study, known as the "Coleman Report", analyzed the performance of over 150,000 students and found that socioeconomic status is more important in determining educational outcomes than differences in school resources (*i.e.* per pupil spending). The study also found that socially disadvantaged students profited from schooling in racially mixed classrooms, and served as a catalyst for the (attempted) desegregation in American public schools.

Sociology of family, gender, and sexuality form a broad area of study. The family is being examined as an institution with such issues as the nuclear family, the single parent, and current statistics that indicate less than 50% of couples living together are married. In the era of the "glass ceiling", feminist sociology is a subfield being studied, particularly with respect to power and inequality. Same-sex marriage is now a major political issue.

Sociology of health and illness focuses on the effects and public attitudes toward illnesses, diseases, disabilities and the aging process. In the current political climate, these issues have become a major focus.

Sociology of the Internet has become paramount in many ways. Few developments have had such an immediate social impact around the world. With its pervasiveness it has literally erased the cultural barriers of international borders. Its timeliness has led to almost instant changes in cultures, societies, and governments, i.e., the "Arab Spring" and a restructuring of the Middle East. Will the vast reaches of China and other areas of Asia become next?

Military sociology is the study of the military as a social group. It is a highly specialized subfield with shared interests linked to survival in combat, and values that are more defined and narrow than within civil society. Topics include the willingness to fight, utilization of women, the military-industrial-academic complex, and the institutional structure of military.

Sociology of race and of ethnic relations encompasses the study of racism, residential segregation, and other complex social processes between different racial and ethnic groups.

Sociology of religion concerns the practices, historical backgrounds, universal themes and roles of religion in society. There is particular emphasis on the recurring role of religion in all societies and throughout recorded history. It is distinguished from the philosophy of religion in that sociologists do not set out to assess the validity of religious truth-claims.

Sociology of work examines trends in technological change, work organization, and employment related to changing patterns of inequality in modern societies. When I was in graduate school in the late 1950's, I participated in a program that developed a new concept called "Job Design". The objective was to enlarge the content of repetitive mundane jobs to make them more meaningful to the worker in the modern industrial factory. I was able to apply some of this once I became a manager in industry.

Sociology overlaps with a variety of disciplines such as anthropology, political science, economics, and industrial psychology. The field of social psychology emerged from the many intersections of psychological and social interests. Applied sociology is connected to the professional discipline of social work and the effect of systems (i.e. family, school, community, and laws) on the individual. Social work is generally more focused on practical strategies to alleviate social dysfunction; whereas sociology provides a thorough examination of the root cause of these problems. For example, a sociologist might study *why* a community is plagued with poverty, whereas the applied sociologist would be more focused on practical strategies on *what* needs to be done to alleviate this burden. The social worker would be focused on *action*; implementing these by means of mental health therapy, counseling, advocacy, community organization or community mobilization.

In 2007, The Times Higher Education Guide published a list of the most cited authors of books in the Humanities' (including philosophy and psychology). Seven of the top ten listed were sociologists. [174]

The most highly ranked journals in the field of sociology are Sociological Perspectives, the American Sociological Review, and the *American Journal of Sociology*.

PHYSICS: THE NEXT FIVE CHAPTERS

The next five chapters are about modern physics, and they are interrelated; so, I will group them together, but treat each as a separate subject. The first four chapters are about the four forces of nature: gravity, electromagnetism, nuclear, and the weak force:

- Chapter 20 is about gravity, as it is presented in Einstein's Theory of Relativity.
- Chapter 21 is about the electromagnetic force. This force gives rise to electricity and magnetism; so is one of the most useful in our daily lives -- even though the theory is difficult to understand.
- Chapter 22 is about the nuclear force, which gives us nuclear power and the atomic bomb.
- Chapter 23 is about the weak force. Except for reading about it in this chapter, you may never be aware it exists.
- Chapter 24 reaches beyond the known physics of today and discusses other possible theories, and where we are currently doing our research.

When I was in graduate school at U.C. Berkeley, I often attended noon luncheons that were open to the entire student body at which prominent people talked, including several of the Nobel Laureates then on campus. Three of my favorites were Glenn Seaborg and Edwin McMillan (who received the Nobel Prize for the discovery of plutonium that became a major component of the atomic bomb) and Edwin Teller (the "Father of the Hydrogen Bomb"). They were some of the foremost physicists of their day; yet, they had the ability to present complex subjects in simple language that all of us students could understand.

In the following chapters, I have attempted to accomplish the same thing. I will be disappointed if you decide to skip over anything. Let me claim that even without a high school education in science, every reader will be able to pass the written test. As Mitt Romney would say: "I'll bet you $10,000."

EINSTEIN AND GRAVITY

Everyone is familiar with Einstein as "The Father of Modern Physics". He had such a profound impact on our understanding of the physical world that it is difficult to know where to begin to discuss his accomplishments. His concept of gravity is near the top of the list.

Einstein was born in 1879 in Germany; later worked in the Swiss Patent Office in Bern, Switzerland; immigrated to America from Germany in 1933 when Hitler came to power; and became a naturalized American citizen. [175] He married twice and had two sons by his first wife. One of his sons, Hans Albert Einstein, followed in his father's footsteps and obtained a Doctor of Technology degree. He immigrated to America at the same time as his father, eventually becoming an Associate Professor of Hydraulic Engineering at the University of California, Berkeley. [176] He was on the engineering staff at the same time I was a Teaching Assistant. I would often encounter him in the Engineering building.

That is enough personal details about the Einstein family; now on to his scientific resume.

After Einstein graduated with a degree in physics and was unable to get employment elsewhere, he went to work for the Swiss Patent Office evaluating applications for patents for electromagnetic devices. As he analyzed questions related to the transmission of electrical signals and electrical-mechanical synchronization of time, this extended his knowledge into the practical world of physics and became an important backdrop for his later scientific adventures. This is apparent with his "thought experiments" that eventually led Einstein to his radical conclusions about the nature of light and the fundamental connection between space and time. [177]

In 1905 he published four groundbreaking papers on the following subjects: photoelectric effect, Brownian motion, relativity, and the equivalence of matter and energy. These brought him to the notice of the academic world." [178] In 1921, Einstein was awarded the Nobel Prize in Physics. Because his theory of relativity was still controversial, the Nobel Prize was officially bestowed for his prior explanation of the photoelectric effect.

PHOTOELECTRIC EFFECT:

What is the photoelectric effect? It is a phenomena first identified by Hertz a dozen years earlier, in which electrons are emitted from a surface as a consequence of absorbing energy. These are called photoelectrons. For example, an iron that is heated to red hot emits photoelectrons seen as emitted light. An earlier explanation for these emissions was based on a wave theory of light promulgated by the scientist Maxwell, but there were problems with the theory that could not be reconciled. Einstein solved the paradox by describing light as composed of discrete quanta, now called photons, rather than continuous waves, and he provided a formula that measured the energy in each quantum of light. When this energy is greater than a certain threshold level, an electron is ejected. While this may not seem significant to a lay-person, this discovery led to the quantum revolution in physics. [179]

Einstein's theory of light quanta was nearly universally rejected by all the leading physicist of his day. It was not accepted until fifteen years later after an experiment by Millikan. In my physics lab in college, we students had to repeat the Millikan experiment. I possessed only a marginal ability to accomplish this, and like most other students I got the "phi bate notes" and jobbed my report. As I recall, I got an A. The lab exercise gave us students the self-confidence to call ourselves scientists.

The photoelectric effect helped sell to the scientific community the concept of the dualistic nature of light in which it possesses the characteristics of both waves and particles. This paradox of the dual nature of light as both a particle and a wave remains today without a completely satisfactory explanation.

When I was an undergraduate student, I had a part-time job working for Dr. Julian Blair, Professor of Physics at the University of Colorado, and his specialty was in light, optics, and photography. I was mostly a "go-fer" and contributed almost nothing to the advancement of science at that time. Perhaps I can now solve this paradox of particles - vs - wave as I peruse the sparks flying from my BBQ grill.

BROWNIAN MOTION:

This phenomenon is seemingly the random drifting of particles suspended in a liquid or gas and is named after the scientist Robert Brown. Why does this drifting occur? It was Einstein who brought the solution to the problem to the attention of physics and presented it in a way to indirectly confirm the existence of atoms and molecules. The movement occurred because of the continuous excitation of these primary particles in their atomic dance. [180]

THEORY OF RELATIVE MOTION:

Einstein's theories about gravity began with what has been described as "thought experiments." Rather than spending time in a laboratory or playing with formulas, he reasoned with a verbal discussion of physical concepts. It is only after he became satisfied with his reasoning that he developed the mathematics necessary to prove the theory. This process is described in the book he wrote five years before he was presented with the Nobel Prize. In the book, he addresses you person-to-person and invites you to join him in his intellectual pursuit. The simplicity of the approach is disarming, and it is amazing where it will eventually lead. The following is Einstein's Chapter Three: <u>Space and Time in Classical Mechanics</u>.

> *"The purpose of mechanics is to describe how bodies change their position in space with "time." I should load my conscience with grave sins against the sacred spirit of lucidity were I to formulate the aims of mechanics in this way, without serious reflection and detailed explanations. Let us proceed to disclose these sins.*
>
> *"It is not clear what is to be understood here by "position" and "space." I stand at the window of a railway carriage which is travelling uniformly, and drop a stone on the embankment,*

without throwing it. Disregarding the influence of the air resistance, I see the stone descend in a straight line. A pedestrian who observes the misdeed from the footpath notices that the stone fall to earth in a parabolic curve. I now ask: Do the "positions" traversed by the stone line "in reality" on a straight line or on a parabola? Moreover, what is meant here by motion "in space"? From the considerations of the previous section the answer is self-evident. In the first place we entirely shun the vague word "space," of which, we must honestly acknowledge, we cannot form the slightest conception, and we replace it by "motion relative to a practically rigid body of reference." The positions relative to the body of reference (railway carriage or embankment) have already been defined in detail in the preceding section. If instead of "body of reference" we insert "system of co-ordinates," which is a useful idea for mathematical description, we are in a position to say: The stone traverses a straight line relative to a system of co-ordinates rigidly attached to the carriage, but relative to a system of co-ordinates rigidly attached to the ground (embankment) it describes a parabola. With the aid of this example it is clearly seen that there is no such thing as an independently existing trajectory ("path-curve"), but only a trajectory relative to a particular body of reference.

"In order to have a complete description of the motion, we must specify how the body alters it position with time: i.e. for every point on the trajectory it must be stated at what time the body is situated there. These data must be supplemented by such a definition of time that, in virtue of this definition, these time-values can be regarded essentially as magnitudes (results of measurements) capable of observation. If we take our stand on the ground of classical mechanics, we can satisfy this requirement for illustration in the following manner. We imagine two clocks of identical construction; the man at the railway carriage window is holding one of them and the man on the footpath the other. Each of the observers determines the position on his own reference-body occupied by the stone at each tick of the clock he is holding in his hand. In this

> *connection we have not taken account of the inaccuracy involved by the finiteness of the velocity of propagation of light. With this and with a second difficulty prevailing here we shall have to deal in detail later."* [181]

From this above simple but eloquent language, Einstein then proceeded to establish the concept that in space there are no fixed coordinates and all motion is relative. He later proceeded to add the mathematical formula that supported the theory. In his theory, Einstein states that all uniform motion is relative, and there is no absolute state of rest. [182] In our modern world in which space craft travel to the far reaches of the universe, we now accept that there is no "ground zero" at rest anywhere in the universe since everything is in motion. But Einstein went further and stated that this phenomenon of relative motion applies to all the laws of physics including both the laws of mechanics and electrodynamics. This incorporates the principle that the speed of light is the same for all observers throughout the universe.

This theory has a wide range of consequences which have been experimentally verified, including such things as the length contraction in space (in time travel, a yard stick can lengthen to 37 inches); time dilation; and relativity of simultaneity (things happening at the same time); and contradicting the notion that the duration of the time interval between two events is the same for all observers. [183]

When Einstein first stated this theory, it set the world of physics upside down and was extremely controversial since it challenged intuition and long-held teachings. The Nobel Committee felt Einstein deserved a Nobel Prize, but decided to award it for his work on the Photoelectric Effect and avoid the controversy surrounding Relativity.

MATTER ENERGY EQUIVALENCE:

Most people today in our nuclear age are familiar with the equation $E=MC$. What does that mean? It means that the mass of a body is also a measure of its energy content -- mass and energy are equivalent. A hundred years later after he proposed his theory, we are all familiar with the atomic bomb in which a small mass of Plutonium is converted to energy with a gigantic release of energy in an atomic explosion. [184]

The equation contains C^2, which is the constant that gives the ratio of energy to mass. C is the expression physicist's use for the speed of light (in a vacuum). It is the maximum speed at which all energy, matter, and information throughout the universe can travel, and is predicted by current theory to be the speed of gravity (that is, the speed of gravitational waves -- if they exist). Such particles travel at the speed of C, which is 186,282 miles/second. [185]

The Theory of Relativity inter-relates space and time. Many of the foremost physicists of his day had difficulty in accepting Einstein's concept and it remained controversial for many years.

EINSTEIN'S THEORY OF GENERAL RELATIVITY:

The Theory that was published in 1905 was limited to the special case in which all uniform motion is relative and there is no absolute state of rest, but Einstein realized there were other phenomena not yet explained. In 1907 he came up with a further concept, but the mathematics was so tricky that the Theory of General Relativity was not perfected until 1915. As a German scientist he published it in his book. The announcement of the theory by a German landed with a dud on the scientific community. World War 1 in which the Germans were fighting against the English was raging at the time and English speaking countries scorned all things German -- including Einstein, who still lived in Germany. How dare that German chap with his new theory of gravitation upstage our great Englishman, Isaac Newton, who discovered gravity! [186]

A centerpiece of the Einstein's theory is gravitation, but there are two different theories explaining gravity: the earlier theory in 1687 by Isaac Newton, and this more recent one in 1917 by Albert Einstein.

Newton explains gravity as an "attractive" force between two masses: an apple falls downward from a tree because the mass of the apple is attracted by the pull of the earth. For all practical purposes, the theory of Newton works until one gets in outer space and encounters masses going at great speeds. For example, it failed to accurately predict the orbit of the planet Mercury as it approached the sun and sped up during its orbit. Newton's law works fine here on earth and gives good answers, but in outer space where bodies can approach the speed of light it was an erroneous concept.

In his theory of General Relativity, Einstein explains that the apparent gravitational attraction results from the "warping of space-time by these masses. It is a concept that space-time is distorted by matter. Free-falling objects in space appear to be moving along straight paths, but are actually following the warped contours of space-time. It is as if a blanket (space-time) holds a bowling ball (the earth), and the ball pushes the blanket downward in the center, warping it. This contour of the blanket is the warped space-time. Gravity is a consequence that empty space is not perfectly "flat", as we might intuitively think, but is deformed where a mass occurs, and gravity is result of the "warping of space-time" by this mass. This concept is not entirely original with Einstein. The idea of a unified space-time is stated by Edgar Allan Poe in his essay on cosmology titled *Eureka* (1848) that *"Space and time are one."* In 1895 in his novel *The Time Machine*, H.G. Wells wrote, *"There is no difference between time and any of the three dimensions of space except that our consciousness moves along it, and any real body must have extension in four directions: it must have Length, Breadth, Thickness, and Duration."* [187]

The effects of gravitation are ascribed by Einstein to space-time curvature instead to an unexplained force such as Newton's "attraction" between two masses. A huge mass, such as the earth with its own relative motion travel through space, causes space-time to become warped. Another mass, such as the apple following its own trajectory though space, will have its own space-time warp. The result is an (apparent) force generated because of the relative motion between the two masses -- a force we call gravity. So the apple falls from the tree not because it is attracted to the earth, but because of the space-time warp of the apple and that of the earth as they both travel through space on their different trajectories of relative motion. (Yes, an apple falling from a tree is actually in space; so the apple and earth have motion relative to each other.) [188]

Have you ever wished you could fly through space? Actually you already are. When not attached to the ground, your body exists with its own "warp" as travels through space along its own space-time continuum. Relative motion exists between your body free in space and that of the earth. When your feet are not anchored to the ground, you are like a spacecraft in outer space.

Further from Einstein' book as another illustration of a cause of gravity, in which he invites us to:

> *"Visualize a man living in a big chest that's drifting in empty space. He must tie himself to the floor if he is not to float about. Unable to invoke a space rocket to propel the box, he imagines a "being" pulling on a rope attached to the top*

of the chest and imparting a steady acceleration. The man in the chest can then think himself at home on earth. He no longer tends to float, and any object he releases will fall to the floor. The steady acceleration through empty space will feel to him just like gravity; hence gravity is due to relative motion in space." [189]

So, gravity is not due to "attraction" between masses as Newton proposed, but caused by an acceleration through empty space.

Einstein's General Relativity has other implications beyond gravity, and I will mention only three:

(1) It implies the existence of black holes -- regions in space in which space and time are distorted in such a way that nothing, not even light, can escape, which becomes an end-state for massive stars.
(2) The bending of light by gravity can lead to the phenomenon of gravitational lens-ing, where multiple images of the same distant object are visible in the sky. Light is bent when traveling near a huge mass.
(3) General Relativity also predicts the existence of gravitational waves, which have (allegedly) since been measured indirectly.

Electromagnetic radiation is discussed in the next chapter. Darn, I wish I was still a college student and had the opportunity to explore these scientific challenges and help find that "unified Field". On the other hand, I would not want the stress of all those student loans hanging around my neck after graduation like an albatross. In these modern times, knowledge is gained by students with a considerable financial penalty.

ELECTROMAGNETISM

Electromagnetism is the science concerned with the forces that occur between electrically charged particles. It is one of the four fundamental forces in nature, the other three being gravitation, the strong force, and the weak force.

Electromagnetism is present as both an electric field and a magnetic field. Both fields are simply different aspects of electromagnetism and intrinsically related; thus, an electric field generates a magnetic field, and conversely a changing magnetic field generates an electric field. When electric current flows down a wire, a magnetic field exists around all sides of the wire. The effect is reciprocal; a current exerts a force on a magnet, and a magnetic field exerts a force on a current. This relationship between magnetic fields and currents is extremely important, because it is the phenomenon that gives us electric motors. [190]

Now with that brief comment about electromagnetism, let's move to the big picture and place it within the physics of fundamental forces. There are four forms of energy in our universe, each of which exerts a force. They are:

- *Gravitation*: first identified by Newton and later explained by Einstein as being a consequence of the relative motion of all things in the universe. (It was discussed in the last chapter.)
- *Electromagnetism*: This includes electricity, and a spectrum

of radiation that includes radio waves, microwaves, infrared radiation, visible light, ultraviolet radiation, X Rays, and gamma rays. (It is the subject of this chapter.)
- *Strong Nuclear Force*: that is involved with atomic bombs and nuclear energy. (It will be discussed in a later chapter.)
- *Weak Force*: is responsible for radioactive decay of subatomic particles, but we do not normally encounter it in our daily lives. (It will be discussed in a later chapter.)

We think of our universe as operating as a common entity, but physicists have been unable to reconcile these four forms of energy in our universe with each other. Einstein spent the last decades of his life at this mission, but failed to develop a theory of a "Unified Field" that would explain gravitation with the other three forces.

In this chapter we will address electromagnetic radiation (EMR). The history of electromagnetism is complicated by the tangled way in which it developed. Investigations began with early pioneers such as Faraday, Gauss, Ampere, and others over several centuries. Their concepts were finally pulled together in 1864 by James Clerk Maxwell, a Scottish physicist from Edinburgh. Standing on the shoulders of those who came before, Maxwell was able to develop a set of mathematical equations that showed the interrelation of the many variables involving electricity and magnetism. A hundred-fifty years after Maxwell's work, some of the concepts are not yet fully understood by scientists. [191]

Electromagnetic radiation (EMR) exhibits wave-like behavior as it travels through space. The spectrum of frequencies within these waves includes radio waves, microwaves, infrared radiation, visible light, ultraviolet radiation, X-rays, and gamma rays. The form of EMR we are most familiar with is visible light; however, radio waves give us TV and radio, the microwave oven works with microwaves, we get sunburned with ultraviolet waves, and we view our bones and innards with X-rays. So EMR is an important phenomenon in our daily lives.

It was left to Maxwell to pull together a summary of what was then known and present a unified concept: the Electromagnetic Theory. "*His work in producing a unified model of electromagnetism is one of the greatest advances in physics. ... His contributions to science are considered by many to be of the same magnitude as those of Isaac Newton and Albert Einstein.*" [192]

What was Maxwell's concept to explain this new phenomena,

the electromagnetic force? He showed that the equations predict the existence of waves of oscillating electric and magnetic fields that travel through empty space at a speed that could be predicted from simple electrical experiments. Maxwell wrote:

> "The agreement of the results seems to show that light and magnetism are affections of the same substance, and that light is an electromagnetic disturbance propagated through the field according to electromagnetic laws."

The following paragraphs contain a few comments about each of the phenomena within the EMR spectrum, and an appendix contains the entire spectrum of frequencies.

RADIO WAVES are an EMR with frequencies lower than infrared light. Naturally occurring radio waves are made by lightning and produce static on our radio. Generated radio waves are used for radio communication, broadcasting, radar, satellite communication, computer networks and innumerable other applications. If you are a teenager, your text messages are carried by radio waves (even if you are not a teenager -- radio waves are an equal opportunity provider). Different frequencies of radio waves have different propagation characteristics in the earth's atmosphere; long waves may cover a part of the earth very consistently, shorter waves can reflect off the ionosphere and travel around the world, and much shorter wavelengths reflect very little and travel on a line of sight. All EMR waves travel at the speed of light. [193]

MICROWAVES are EMR with frequencies longer than radio waves. Microwave ovens may be described as "microwave", but the name indicates only that their waves are "small" and not necessarily in the microwave range. [194]

INFRARED WAVES are EMR that include most of the thermal radiation emitted by objects near room temperature. Infrared light is used in industrial, scientific, and medical applications. Night-vision devices using infrared illumination allow people or animals to be observed without detection. Infrared imaging cameras are used to detect heat loss in insulated systems, observe changing blood flow in the skin, and overheating of electrical apparatus. Much of the energy from the Sun arrives on Earth in the form of infrared radiation, and the balance between absorbed and emitted infrared radiation has a critical effect on the Earth's climate.

VISIBLE SPECTRUM - LIGHT: is the EMR that is visible to the human eye, which excites the pigment molecules in the retina. A human eye will see wavelengths from about 400 nm (purple) to 700 (red) and has its maximum sensitivity at around 540 nm in the green region of the optical spectrum. The spectrum does not, however, contain all the colors that the human eyes and brain can distinguish. Unsaturated colors such as pink can be seen, but are absent from the optical spectrum, because they can be made only by a mix of multiple wavelengths.

Many species can see light with frequencies outside the human "visible spectrum". Bees and many other insects can see light in the ultraviolet, which helps them find nectar in flowers. Plant species that depend on insect pollination may owe reproductive success to their appearance in ultraviolet light, rather than how colorful they appear to humans. Birds, too, can see into the ultraviolet, and some have sex-dependent markings on their plumage that are visible only in the ultraviolet range. [195]

ULTRAVIOLET WAVES are EMR with a frequency higher than that of visible light. It is named because the spectrum consists of frequencies higher than those that humans identify as the color violet. These are invisible to humans, but visible to a number of insects. UV light is found in sunlight and specialized lights such as black lights. It can cause chemical reactions, and causes many substances to glow or fluoresce. However, the entire spectrum of ultraviolet radiation has some of the biological features of ionizing radiation, in doing far more damage to many molecules in biological systems than is accounted for by simple heating effects (an example is sunburn). Although ultraviolet radiation is invisible to the human eye, most people are aware of the effects of UV through sunburn. Ultraviolet is also responsible for the formation of vitamin D in all organisms that make this vitamin (including humans). The UV spectrum thus has many effects, both beneficial and damaging, to human health. [196]

X-RAY is an EMR with waves of higher frequencies than UV rays. In many languages, it is called Röntgen radiation because Röntgen was its discoverer and named it X-radiation to signify an unknown type of radiation. X-rays can penetrate solid objects, and their most common use is to take images of the inside of objects in diagnostic radiography and crystallography. [197]

GAMMA RAYS are EMR of high frequency (very short wavelength). Gamma rays are produced on Earth by decay of high energy states in atomic nuclei (gamma decay). Gamma rays are ionizing radiation and are thus biologically hazardous. [198]

So, that is the spectrum of electromagnetic radiation. As discussed in a previous chapter, when Albert Einstein was working in the patent office, he published his paper on the photoelectric effect. This stated that light interacts with matter as discrete "packets" (quanta) of energy, which seemed to contradict the existing wave theories of light at that time. So; which is it? Is electromagnetic radiation something transmitted as a ware, or is it transmitted as packets of photons? The answer is both; but don't ask physicist to explain that dichotomy with a concept that we mortals can grasp.

SUMMARY: Permit me to summarize what we have learned about Electromagnetism, one of the four fundamental forces within our universe. We have learned how to utilize tools such as electricity, radio waves, and X-Rays to improve our "quality of life", but scientists have a lot to learn about fundamental aspects of electromagnetism. While some scientists think these concepts have already been established, others (such as my humble self) think the definitive research has not yet been successful. (Where do EMR waves fit within a "unified field", and what is the nature of those particles -- are they consistent with string theory?) We still have a lot to learn, and a reader has cause to remain somewhat confused and puzzled about electromagnetic radiation.

THE NUCLEAR BOMB

Why is an atomic bomb so powerful and where does that energy come from? The answer is from the Strong Nuclear Force, which is the potential energy that binds the protons and neutrons together inside the nucleus of the atom. When the nucleus of the atom is "split" apart, that potential energy is released as the kinetic energy of the explosion. Because of mass–energy equivalence (i.e. Einstein's famous formula $E = mc^2$), a huge amount of energy is released in a chain reaction when twenty pounds of plutonium is converted into energy, hence the huge explosion

Perhaps we should start with a brief review of the atom. Bear in mind that no one has ever seen an atom, let alone any of its constituent parts. Scientist think they exist because of many experiments that have established their validity with 99+.99% probability. Is that good enough for you? It is for me.

An atom consists of a nucleus that contains protons and neutrons and the electrons whirling in orbit around the nucleus. The Helium atom has 2 electrons in orbit around a nucleus containing two protons along with either one or two neutrons, depending on the isotope. Plutonium-239 (used in the atomic bomb) has 93 protons and 146 neutrons in the nucleus. When the nucleus of Plutonium is bombarded by neutrons, it releases gamma radiation plus more neutrons, and those neutrons can sustain a nuclear chain reaction -- hence the bomb.

This Strong Nuclear Force is one of the four fundamental forces of nature we have previously listed: Electromagnetism, Gravity, and the Weak Force. It is 100 times stronger than Electromagnetism and thousands of times stronger than Gravity and the Weak Force.

A bit complicated? Read the following from a reference:

> "In 1934, The neutron was discovered, and this revealed that atomic nuclei were made of protons and neutrons, held together by an attractive force. The nuclear force was conceived to be transmitted by particles called mesons .... Further understanding revealed these mesons to be combinations of quarks and gluons. ... This new model allowed the strong forces that held nucleons together, to be felt in neighboring nucleons, as residual strong forces." [199]
> (You don't have to memorize all that stuff about mesons, quarks, and gluons -- there will be no written test.)

The nuclear force has been at the heart of physics ever 1932 with the discovery of the neutron. It was a big task to describe the nuclear force phenomenologically (17 letters and another long word I love to see, but will probably never use again). There has been substantial progress in experiment and theory related to the nuclear force with most basic questions settled in the 1960s and 1970s. (This is a biased position taken by scientists, but not supported by actual fact -- there are fundamental questions yet to be learned such as what is its relationship to Electromagnetism?)

I will now discuss the subject of construction of an atomic bomb. While I had a top secret clearance sixty years ago when I was a naval officer, I have never been privy to any secret information about the bomb and will only tell what I have read in scientific journals or on the internet.

A chain reaction is started with an explosion that brings together a critical mass of fissionable material such as Plutonium 239 or Uranium 235. A critical mass is the smallest amount of fissile material needed for a sustained nuclear chain reaction. For Plutonium 239 the critical mass is about 10 kg, and for Uranium 235 it is about 52 kg, according to the reference material. [200] The critical mass depends upon its nuclear properties, its density, its shape, its enrichment, its purity, and its temperature. The critical mass must be brought together quickly before the explosion tears the bomb apart. There are two known methods for doing this. In a gun-type bomb, portions of the fissionable material are combined into a critical mass by shooting one part toward the other. In an implosion weapon, the material is kept apart from a critical mass with cushioning material, and then an explosion on the periphery compresses it together into a critical mass.

These two types of bombs were designed during the Manhattan Project of World War Two at Los Alamos under American physicist Robert Oppenheimer. A test implosion weapon was detonated on July 16, 1945, near Alamogordo, New Mexico. The Hiroshima bomb, known as Little Boy, was a gun-type fission weapon made with uranium-235, a rare isotope of uranium extracted in factories in Oak Ridge, Tennessee. The Nagasaki bomb, Fat Man, was an implosion-type nuclear weapon using plutonium-239, a synthetic element created in nuclear reactors at Hanford, Washington. [201] Since the end of the war, American nuclear bombs have been designed and built at the Livermore Lawrence Nuclear Laboratory, which is administered by the University of California. After I obtained my master degree at U.C. Berkeley, I resisted the urge by some professors for me to go to work at the Livermore Laboratory.

Since the bombings of Hiroshima and Nagasaki, nuclear weapons have been detonated on over two thousand occasions for test purposes. The only countries that acknowledge possessing such weapons are the United States, the Soviet Union, the United Kingdom, France, the People's Republic of China, India, Pakistan, and North Korea. In addition, Israel is also widely believed to possess a couple hundred nuclear weapons, though it does not acknowledge having them. The United States and Russia each have about 5000 nuclear weapons, which amounts to about 95% of all those in the world.

Another type of an even bigger nuclear weapon, called the hydrogen bomb, produces its energy through nuclear fusion. It relies on fusion reactions between isotopes of hydrogen (deuterium and tritium). However, all such weapons derive a significant portion of their energy from fission. This is because a fission weapon is required as a "trigger" for the fusion reactions, and the fusion reactions can then trigger additional fission reactions.

I participated in the first H bomb test held in 1954 in the South Pacific near the island of Kwajalein. I was in my stateroom aboard a destroyer escort located a hundred miles from the blast when I heard the loud "swish" -- the sound wave that had carried across the water for all that distance. My ship was dispatched to a nearby island to pick up natives who were exposed to radioactive fallout from a passing cloud to decontaminate them. The crew of a Japanese fishing boat was also exposed. The incident caused widespread concern around the world regarding atmospheric nuclear testing, and "provided a decisive impetus for the emergence of the anti-nuclear weapons movement in many countries" [202]

In a previous book I have recounted additional details about that nuclear decontamination process. A navy doctor and I were two officers who were members of a division staff and not part of the ship's crew. He was designated as officer-in-charge of the process and assigned me with the task to utilize a hand-held Geiger counter and carefully scan each of the naked South Sea maidens as they exited the shower to insure they were adequately decontaminated. We were all set to implement our mission, but as we entered the harbor saw that another naval vessel was already taking the natives aboard. Our mission had been preempted so my virtue remained intact. Our ship returned to Kwajalein. A week later I passed by the barracks where the natives were now housed. Peering out through a web-wire fence were the native girls -- and not a beauty among them. Where was Dorothy Lamour?

Nuclear power was in my shadow because of my academic background in engineering physics. As a naval officer I was pressured to go into nuclear submarines; I declined, got out of the navy, and went to graduate school at U.C. Berkeley instead. A professor there urged me to pursue a career in nuclear; I declined and instead spent my career as an industrial manager making glass bottles. While manufacturing glass bottles held little glamour, fortunately, they were not radioactive.

NUCLEAR WEAK FORCE

No one except a scientist would have any interest in the nuclear weak force. It does not enter into our daily life process.

The weak nuclear force is one of the four fundamental forces of nature, alongside the strong nuclear force, electromagnetism, and gravity. It is responsible for the radioactive decay of subatomic particles and initiates the process known as hydrogen fusion in stars.

It is termed *weak* because it is several orders of magnitude less than that of the other forces. The weak force was originally described in the 1930s by Fermi. In 1968, the electromagnetic force and the weak interaction were unified, when they were shown to be two aspects of a single force, now termed the electro-weak force. [203]

Did you understand that? If so, please contact the nearest physicist and explain it to him (or her). Actually, much of the phenomena surrounding the nuclear weak force can be dealt with only with quantum mechanics, which is the mathematical exercise physicist utilize that seems to give results even though they are unable to describe it with an understandable concept (even explain it to themselves). More about quantum mechanics in the next chapter.

While the Nuclear Weak Force languished in the back waters of physics for decades, suddenly in 2011 it has arrived front and center in current theory. About fifty years ago, a young researcher at Edinburgh University, named Higgs, asked himself where mass comes from. (Where the Mass comes from -- he obviously was not Catholic.) Like the tendency of an apple to fall to the ground, the existence of mass did not seem to need a formal explanation. Higgs went to work on the question and came up with an answer; particles have mass because of their interaction with a previously unknown field that permeates space. The field now bears his name as the "Higgs Field". It is needed to explain a phenomenon called electroweak symmetry breaking, which divides two of the fundamental forces of nature, electromagnetism and the nuclear weak force. When that division happens, quantum mechanics predicts the existence of a new particle that has become known as the Higgs bosom. Wow! The big excitement came in December 2011 when its possible discovery was announced with research from the Swiss LDH (Large Hadron Collider). [204]

Since the nuclear weak force falls in the category of "research", a further discussion will be deferred until the next chapter.

SCIENTIFIC RESEARCH

I have had a lifelong interest in science as I watched scientists push out the frontiers of knowledge. In graduate engineering school at the University of California, Berkeley, I conducted research related to the manufacture and inspection of products by personnel working on factory conveyor lines. Issues involved line speed, detection ability, fatigue, operator training, and the utilization of special devices. All these things were pragmatic considerations in the industrial environment. The principle contribution of my research was to qualify me for an advanced degree, and an invitation to join the Sigma Xi Scientific Honorary, the prestigious institution for scientists. Our Berkeley chapter included a number of Nobel Laureates; I was at the bottom of the totem pole.

Sigma Xi publishes a journal several times a year of all the latest research findings; so devouring this information for the past sixty years has enabled me to keep abreast with what is going on. In the following paragraphs, I will outline where current research activities are focused in the physical sciences.

UNIFIED ENERGY FIELD

We are still pursuing the questions addressed by Galileo and other pioneers six centuries ago: what is the nature of our universe? This endeavor focuses on two types of activities: what are the smallest particles that go together to make up the universe, and how do these interact to create a unified energy field that ties gravitation to the other three forces of nature: electromagnetic, nuclear, and the weak force.

Two centuries ago as we looked microscopically, we thought the atom was the smallest unit of matter. In the intervening years we kept finding smaller and smaller particles such as electrons, protons, neutrons, quarks, mesons, Higgs bosons, etc. A century ago a lone scientist, named Einstein, working a part-time job to pay the housekeeping bills, performed "thought experiments" at his kitchen table that defined an entirely new concept of the universe, named the Theory of Relativity. Today we are still trying to reconcile that concept together with all the particles we have now found with some sort of rational unified field.

The center of this research is currently taking place at the Swiss LHC (Large Hadron Collider) where particles are being blasted apart in a search for even smaller particles, hoping to eventually identify the smallest unit of matter, or a unified field. Whether this will be just "nice to know" information or lead to new applications to improve our "quality of life" remains to be seen. It is relatively inexpensive research to conduct and provides employment for a large scientific community.

There are a number of other scientific investigations currently being pursued, which include: Quantum Mechanics, String Theory, and Dark Matter.

QUANTUM MECHANICS:

Quantum Mechanics is a body of scientific principles that explains the behavior of matter and its interactions with energy on the microscopic scale of atoms and atomic particles. [205] We have utilized the mathematics for several decades and still do not have any concept to explain why quantum mechanics gives good answers. It is like trying to understand geometry without the use of triangles, circles or squares. Some aspects can seem counter-intuitive because they describe behavior quite different than what we can conceptualize, and in the words of Richard Feynman, "*Quantum Mechanics deals with 'nature as she is -- absurd.*" It provides a useful tool to give us empirical answers, but has provided us with no physical concept we can visualize.

STRING THEORY:

We have only one universe so intuitively one would think that the fundamental Laws of Nature would reign supreme throughout. If that were true, then why are we unable to reconcile the laws that govern gravitation with the other three energy forces? Gravitation can be explained by Einstein's Theory of Relativity. Why can't Electromagnetic waves find some compatible explanation?

String theory is an attempt to reconcile quantum mechanics and gravity. It postulates that the smallest constituents of matter are not subatomic particles like the electron but extremely tiny one-dimensional strings of energy. These elemental strings can vibrate at different frequencies, like the strings of a violin, and the different modes of vibration correspond to different fundamental particles and forces. [206]

String theory believes that the electrons and quarks are not 0-dimensional objects (operating at a point), but rather 1-dimensional oscillating lines ("strings"). They can oscillate, giving the particles their charge, mass, and spin. String theories require the existence of several extra dimensions to the universe in addition to the four known space time dimensions. [207]

Based on what scientists now know (or think they know), they are left with the dilemma that they have not yet discovered the fundamental particle (or whatever) that is at the bottom of the particle pile, or metaphorically is at the "top of the particle food chain." What feeds the building blocks for quarks that are the building blocks for nuclei of electrons, which combine to form molecules, which become the material for all matter (including us)?

DARK MATTER:

Dark matter is another area of research currently receiving much attention. It is matter that neither emits nor scatters light or other electromagnetic radiation, and so cannot be directly detected via optical or radio astronomy. Dark matter is believed to constitute 83% of the matter in the universe and 23% of the mass-energy. It was postulated in 1934 to account for evidence of "missing mass" in the orbital velocities of galaxies in clusters. Based on new evidence of the accelerating rate of expansion, it constitutes a whopping three quarters of the total energy of the universe, and is the invisible elephant in the room of science. [208] Though the existence of dark matter is generally accepted by the mainstream scientific community, some alternative theories have been proposed to explain the anomalies that dark matter is intended to solve. Will Dark Matter have the same ending as the "ether", the existence of which was discounted several decades ago, or will it instead help explain some new concept of a unified field theory?

THE STANDARD MODEL:

The Standard Model of particle physics is a theory concerning the electromagnetic, weak, and strong nuclear interactions together with known subatomic particles. Because of its success in explaining a wide variety of experimental results, the Standard Model is sometimes regarded as a theory of almost everything. Still, it falls short of being a complete theory of fundamental interactions because it does not incorporate the physics of dark energy nor theory of gravitation as described by general relativity. Nevertheless, the Standard Model is important to theoretical and experimental particle physicists alike. [209]

Anyone who reads scientific journals is frequently subjected to headlines announcing the discovery of some new elementary particle. Let me place this subject in perspective.

During the 20th century, physicists have concentrated great effort and resources on the discovery of elementary particles. In 1907, Ernest Rutherford discovered the atomic nucleus which contained most of the mass of the atom. An atom is orders of magnitude too minute to be seen by any scientific instruments, hence we know of its existence only because of numerous experiments. The nucleus is tens of thousands of times even smaller than the distance from it to the closest electron orbiting the perimeter of the atom. So, the atom consists of a minute speck of the nucleus and a gigantic empty space out to the electron where the chemical business of the atom takes place. The nucleus consists of protons and neutrons that are held together with nuclear glue. When the nucleus is split open and the potential energy within it is converted to Kinetic energy, the atomic bomb is the result.

After the discovery of the atom, then a host of elementary particles have been discovered. [210]

In 1947, a particle known as the pi meson was discovered. It came in three varieties with positive, negative, and zero charge. It had some mass. The charged varieties decayed rapidly into mu mesons, heavy electrons and neutrinos. To celebrate the finding, the eminent scientist Edward Teller, the "Father of the H Bomb", who is not generally known for light verse, nicely summed up the sense of surprise at the discovery with the following bit of doggerel:

> There are meson pi, and there are mesons mu.
> The former ones serve us nuclear glue.
> There are mesons tau -- or so we suspect --
> And many more mesons which we can't yet detect.
>> Can't you see them at all?
>> Well, hardly at all.
> For their lifetimes are short
> And their ranges are small.
> The mass may be small, and the
> mass may be large.
> We may find a positive
> or negative charge.
>> And some mesons will never
>> show on a plate,
>> For their charge is zero, though
>> their mass is quite great.
> What, no charge at all?
> No, no charge at all.
> Or if Blackett is right,
> It's exceedingly small. [211]

The antiproton was found in 1933, when it was found that every particle had an antiparticle. The research was conducted in the Bevetron on the campus at the University of California, Berkeley.

The quark was discovered in the 1960's in the Stanford Linear Collider near Palo Alto. Electrons are accelerated to an enormous energy travelling down a two mile straight path. (It crosses under the #760 freeway on which I frequently travel from San Mateo to Palo Alto.) Electrons collide with protons and results show they have struck hard objects inside the proton – now identified as quarks. There are six different "flavors" of quarks: up, down, strange, charmed, bottom, and top. I wonder if any of the quarks escaped the linear accelerator and struck my auto -- or me.

Other elementary particles not mentioned above include bosons, k mesons, fermions, strings, and gravitons, but our palette of elementary particles is already too overflowing to go into these in any detail.

The current priority quest is a hunt for the Higgs boson. The Large Hadron Collider (LHC) located in Switzerland was built primarily to establish the existence of this important elementary particle. In 2012, there is preliminary evidence that it may exist.

Where do all these elementary particles fit in the larger picture of physics and an understanding of our universe? That remains to be seen. New theories abound concerning the existence of dark energy, and additional dimensions in space time beyond the four we already know. Perhaps there exists some "field" throughout the universe in which all these elementary particles arise, or co-exist? Since we do not have a concept to explain the quantum mechanism that we utilize to explore the action of elementary particles, we are not yet able to explain this confusing aspect of physics. Is there a "unified field" that would tie together gravity with the other three: electromagnetic, nuclear, and the weak force? If so, Einstein was unable to find it. Or is our universe just like a bowl of plum pudding, composed of numerous random particles each vibrating with their own musical game? A scientist should not end his quest with a question, but that is the scientific hand we were dealt, so I will end there.

Bon voyage.

PHILOSOPHY
Love of Wisdom

Philosophy spelled backward is yhposolihp -- which is meaningless. The word if spelled from the front end is intimidating, particularly if it is tackled with a grunt and a head bowed in prayer. Philosophy is a heavy subject that needs a light touch and some irreverence -- it is not a religion. It is also not the sum-total of all knowledge, as some would have you think; but only the study of how to tackle a question. So much for my naïve introduction of the philosophy of philosophy.

I am ambivalent about this subject, named by the Greeks over two thousand years ago at the beginning of Western Civilization, when it included most known morsels of wisdom. On the one hand it has served us well as we travelled down the road of history for two millennia; but on the other hand, much of the subject matter has gradually been eroded into science sub-cultures named physics, chemistry, biology, anthropology, and others -- so the field of philosophy has shrunk and is a skeleton of its former self. What is left? [212]

Perhaps I am not qualified to answer since I am not academically trained as a philosopher. However, I am trained as a scientist and the core of philosophy remains the same: utilizing the scientific method to explore basic principles. We want to know that a fact is a fact. "Philosophy is related to science in two ways: it logically precedes science; it also completes it." [213] We can believe the findings of science because they have been subjected to the disciplines of philosophy.

The main areas of study in philosophy include epistemology, logic, metaphysics, ethics and aesthetics. These are:

- *Epistemology* is concerned with the nature of knowledge such as the relationships between truth, belief, and theories of justification.
- *Logic* is the study of the principles of valid inference and correct reasoning. Today the subject of logic has two broad divisions: mathematical and philosophical logic.
- *Metaphysics* is the study of reality such as existence, the relationship between mind and body, and events and causation. A traditional branch is cosmology.
- *Ethics* is concerned with the question of the best way to live, and whether this can be answered. It goes beyond theory into ethical practice. It is also associated with the idea of morality.
- *Aesthetics* deals with beauty, art, enjoyment, perception, and matters of taste and sentiment.

In addition to these main areas of study of philosophy, there are also specialized branches such as the following:

- <u>Philosophy of language</u> explores the nature, the origins, and the use of language.
- <u>Philosophy of law</u> (more commonly called jurisprudence) explores the theories explaining the nature and interpretations of the law in society.
- <u>Philosophy of Mind</u> explores the nature of the mind, and its relationship to the body.
- <u>Philosophy of religion</u>: since religion is founded on "faith" and not on established science, perhaps there is little room in philosophy for religion.
- <u>Philosophy of science</u>: Scientists as philosophers?

Throughout history, many people have built philosophical traditions. Ancient Graeco-Roman philosophy was a period of Western philosophy, starting in the 6th century BC with Plato and Aristotle. Plato is credited as the founder of Western philosophy.[214] Socrates initiated a more focused study on the patterns of reasoning, the nature of the good life, and the concept of justice. This included Socrates' method of inquiry, known as the Socratic Method, which he applied to key moral concepts such as the Good and Justice. To solve a problem, Socrates would break it down into a series of questions that gradually distill the answer a person would seek. This approach led to the present day use of the scientific method, in which hypothesis is the first stage.

The Renaissance ("rebirth") was a period of transition between the Middle Ages and modern thought when a shift in philosophy occurred away from technical studies in metaphysics, and theology towards inquiries into morality. The study of the classics and the humane arts enjoyed a scholarly interest previously unknown in Christendom. New movements developed in Europe with the Protestant Reformation and the decline of feudalism. The gradual centralization of political power in nation-states was echoed in the works of Machiavelli (often described as the first modern political thinker).

The modern era of western philosophy is usually identified with the 18th century and is distinguished by its increasing independence from traditional authorities such as the Church, a new focus on the foundations of knowledge, and the emergence of modern physics out of natural philosophy. Early modern philosophers include Descartes, Berkeley, Galileo, Newton, and Adam Smith. Modern philosophies include the following:

- *Realism*: some things have real existence outside the mind.
- *Rationalism*: a view emphasizing the role of human reason.
- *Skepticism*: an attitude that questions any sort of knowledge.
- *Idealism*: nothing can be known outside of the mind.
- *Pragmatism*: truth that does not depend on personal insight.
- *Existentialism*: thinking begins with the human individual.
- *Structuralism*: analyzing the discourses made possible.
- *Analytic philosophy*: stress of argumentation and semantics.

The ideas conceived by a society have profound repercussions on what actions that society performs. The philosophies of Confucius, Sun Zi, Chanakya, Machiavelli, Leibniz, John Locke, Jean-Jacques Rousseau, Adam Smith, Karl Marx, John Stuart Mill, Mahatma Gandhi, Martin

Luther King Jr. and others -- all have been used to shape and justify governments and their actions.

The philosophy of education (progressive education as championed by John Dewey) has had a profound impact on educational practices. The philosophy of war has had a profound effect on statecraft, international politics, and military strategy. Logic has become important in mathematics and computer engineering. The philosophy of science has affected the nature of scientific investigation (for example, the Skinner's behaviorism affected the method of the American psychological establishment). The philosophies of animal rights examine morality in a world that has non-human occupants to consider. Aesthetics helps interpret discussions of music, literature, and art. In general, the various philosophies strive to provide activities with a deeper understanding of the theoretical or conceptual underpinnings of their fields.

Often philosophy has been seen as an investigation into an area not sufficiently well understood to be its own branch of knowledge. What were once philosophical pursuits have evolved into the modern fields such as psychology, sociology, and economics.

Since I am not a philosopher, it was interesting as I did my research to see the different status given to philosophy by the treatments of the 1960's *Colliers Encyclopedia*, the 2007 *Encyclopedia Britannica*, and the 2011 *Wikipedia Free Encyclo*pedia.

The *Colliers Encyclopedia* treats the subject with reverence, reflecting the high status it enjoyed among academics in the midst of the last century. The 25 page long discussion (among the longest of any subject in the entire encyclopedia) left no stone unturned. It placed philosophy on a high pedestal and well above other subjects such as the sciences. Here from their article:

> "*Philosophy precedes science; it also completes it. The scientist in his daily work is compelled to use many concepts which he cannot stop to examine. Philosophy undertakes to examine these concepts and establish their validity.*" [215]

The *Encyclopedia Britannica* was written nearly fifty years later for a modernized audience and takes a more liberal point of view. Here from their article:

> "It is a paradox that no two philosophers would define philosophy in the same way, for throughout its history it has meant different things. For some it has been a search for the wisdom of life. For others it is an attempt to understand the universe, an examination of man's moral responsibilities, and to fathom the divine intentions. For others it is the enterprise of natural science, or an examination of the values of truth, goodness, and beauty. This gives some idea of its extreme complexity and many-sidedness.
>
> "Part of what makes it difficult to find a consensus among philosophers is that they come to it from different fields, with different interests, and that they have different areas of experience." [216]

Then to the current treatment by the *Wikipedia Free Encyclopedia* that has been prepared by a cross-section of today's peers, which states:

> "Philosophy is distinguished from other ways of addressing problems by its critical, generally systematic approach and its reliance on rational argument."

Since this reflects my understanding of the subject, I truly believe it is the most accurate description of this exceeding complex subject.

SUMMING UP

In my morning newspaper I read an article that related to several of my chapters: electromagnetics, physiology, genetics, chemistry, and scientific research. Here are some highlights from the article.

Light Shed on Bird's Magnetic Navigation

Birds are famously good navigators. Some migrate thousands of miles, flying day and night, even when stars are obscured. And for decades, scientists have known that one navigational skill they employ is an ability to detect variations in the earth's magnetic field.

Now, two researchers at Baylor School of Medicine have solved a key part of that puzzle, identifying cells in a pigeon's brain that record detailed information on the earth's magnetic field, a kind of biological compass. ... They have found cells that are tuned to specific directions of the magnetic field. ... Navigation by magnetism includes several steps: birds have to detect a magnetic field, some part of the brain has to register the information, and another part of the brain must compare the incoming information to a stored map. ... The researchers have identified a group of cells in the brain stem that record the direction and strength of the magnetic field and evidence that this is coming from the inner ear of the pigeon. ... In some birds that hide seeds and return later to their caches with astonishing accuracy, the hippocampus grows and shrinks seasonally, presumably as they map their hiding spots.

This is consistent with a well-known study of London taxi drivers that showed drivers with a mental map of London had a hippocampus larger than drivers without this ability. [217]

While the research mentioned in the article is not "earth shaking", it does illustrate that a basic understanding of the sciences is useful as a companion with the morning cup of coffee

END NOTES

1. "Ptolemy", *Wikipedia*, op. cit.
2. "Johannes Kepler" *Wikipedia*, op. cit.
3. *Ibid.*
4. Ibid.
5. "Epiphany: a sudden manifestation or perception.", *Webster New Collegiate Dictionary*, G.C. Merriam Co., 1977
6. Ibid.
7. "Kepler's Laws of Planetary Motion", *Wikipedia*, op. cit.
8. Alan Lightman, "The Accidental Universe", *Harper's Magazine*, December 2011, pg. 35
9. "Edwin Hubble", *Wikipedia*, op. cit.
10. "Johannes Kepler", *Wikipedia*, op. cit.
11. "Charles Darwin", *Wikipedia*, op. cit.
12. "Charles Darwin", *Encyclopedia Britannica*, 2007 Deluxe Edition, Chicago
13. Ibid.
14. Ibid.
15. Ibid.
16. Ibid.
17. Ibid.
18. Ibid.
19. "Louis Leakey", *Wikipedia*, op. cit.
20. Ibid.
21. "Last Universal Ancestor", *Wikipedia*, op. cit.
22. "Pliopithecus", *Wikipedia*, op. cit.
23. "Human", *Wikipedia*, op. cit.
24. "Radiometric dating", Wikipedia, op. cit.

Radioactive decay: All ordinary matter is made up of combinations of chemical elements, which may exist in different isotopes. Each isotope of an element (called a nuclide) differs in the number of neutrons in the nucleus. At some point in time, an atom of such a nuclide will spontaneously transform into a different nuclide, and this transformation may be accomplished in a number of different ways, including radioactive decay. Atoms of a radioactive nuclide decays exponentially at a rate described by a parameter known as the half-life, usually given in units of years when discussing dating techniques. After one half-life has elapsed, one half of the atoms of the nuclide in question will have decayed into a decay product. Isotopic systems that have been exploited for radiometric dating have half-lives ranging from only about 10 years (e.g., tritium) to over 100 billion years (e.g., Samarium-147).

Accurate radiometric dating generally requires that the parent has a long enough half-life that it will be present in significant amounts at the time of measurement. This normally involves isotope ratio mass spectrometry. The precision of a dating method depends in part on the half-life of the radioactive isotope involved. For instance, carbon-14 has a half-life of 5,730 years. After an organism has been dead for 60,000 years so little carbon-14 is left that accurate dating cannot be established. On the other hand, the concentration of carbon-14 falls off so steeply that the age of relatively young remains can be determined precisely to within a few decades.

[25] "Psychology," *Wikipedia*, op. cit.
[26] Ibid.
[27] Ibid.
[28] Ibid.
[29] "Gestalt" *Wikipedia*, op. cit.
[30] "Psychology", *Wikipedia*, op. cit.
[31] Ibid.
[32] "Psychiatry", Encyclopedia Britannica, 2007 Deluxe Edition, Chicago
[33] Ibid.
[34] Ibid.
[35] Alan Lightman, op. cit. pg 36
[36] Ibid.
[37] Don Lincoln, "Psyching Kids for Physics", *Notre Dame Magazine*, Spring 2011, pg. 12
[38] "Galileo Galilei", *Wikipedia*, op. cit.
[39] "Isaac Newton", *Wikipedia*, op. cit.
[40] Ibid.
[41] "Inverse-Square Law", *Wikipedia*, op. cit.
[42] "Gravitation", *Wikipedia*, op. cit.
[43] Ibid.
[44] Ibid.
[45] "Definitions of Mathematics", *Wikipedia*, op. cit.
[46] Ibid.
[47] Ibid.
[48] "Mathematics", *Wikipedia*, op. cit.
[49] Ibid.
[50] Ibid.
[51] Ibid.
[52] "Deciphering Heredity", *Smithsonian Magazine*, January 2012, Vol. 42, pg. 46
[53] "James D. Watson", *Wikipedia*, op. cit.
[54] "Human Cell", *Wikipedia*, op. cit.
[55] "Introduction to Genetics", *Wikipedia*, op. cit.
[56] Ibid.

57 "Mix, Match, Morph", *National Geographic*, Feb. 2012, pg. 46
58 "Introduction to Genetics", *Wikipedia*, op. cit.
59 "Chromosome", *Wikipedia*, op. cit. A chromosome is an organized structure of DNA and protein found in cells. It is a single piece of coiled DNA containing many genes, regulatory elements and other nucleotide sequences. Chromosomes also contain DNA-bound proteins, which serve to package the DNA and control its functions.
60 Ibid.
61 Ibid.
62 "Marie Curie", *Wikipedia*, op. cit
63 Williams and Ceci, "When Scientists Choose Motherhood," *American Scientist* Magazine, March - April 2012, pg. 138
64 "Chemistry", *Wikipedia*, op. cit
65 Ibid.
66 "Periodic Table", *Wikipedia*, op. cit.
67 "Linus Pauling", *Wikipedia*, op. cit.
68 "Electronics", *Wikipedia*, op. cit.
69 "Vacuum Tube", *Wikipedia*, op. cit.
70 Ibid.
71 "Transistors", *Wikipedia*, op. cit.
72 "Integrated Circuits", *Wikipedia*, op. cit
73 Ibid.
74 "Computers", *Wikipedia*, op. cit.
75 Ibid.
76 "Central Processing Unit (CPU)", *Wikipedia*, op. cit.
77 Ibid.
78 Ibid.
79 "Cloud Computing" *Wikipedia*, op. cit.
80 *Webster's New Collegiate Dictionary*, empirical: "relying on experience or observation alone often without due regard for system and theory."
81 *New Family Medical Guide*, Better Homes and Gardens Books, Meredith Corporation, Des Moines, Iowa, 1982, pg. 700
82 "Human Body", *Wikipedia*, op. cit.
83 "Vascular System", Stanford Hospital and Clinics
84 "Blood Pressure", *Wikipedia*, op. cit.
85 "Vascular system" Stanford Hospital and Clinics, op. cit.
86 "Respiratory System", *Wikipedia*, op. cit.
87 "Immune System", *KidsHealth.org*
88 "Lymph System", *Wikipedia*, op. cit.
89 "Thyroid", *Wikipedia*, op. cit.
90 "Endocine system", KidsHealth.org
91 "Digestive System", Encyclopedia Britannica. 2007, Deluxe Edition, Chicago
92 "Digestive System", *Wikipedia*, op. cit.
93 "Pancreas", *Wikipedia*, op. cit.
94 "Urinary System", *Wikipedia*, op. cit.
95 "Nervous System", *Wikipedia*, op. cit.

96 "The Brain", KidsHealth.org
97 "Integumentary System:" *Wikipedia*, op. cit.
98 Nancy Hellmich, Multiple Strategies Needed to Fight Obesity", *USA TODAY*, 5/9/2012, pg. D3
99 Gary Tabes, "The New Obesity Campaigns Have it all Wrong", *Newsweek*, May 14, 2012, pg. 36
100 Ibid.
101 "Nutrition", *Wikipedia*, op. cit.
102 "Glycerol", *Wikipedia*, op cit. Glycerol is a colorless, odorless, viscous liquid that is widely used in pharmaceutical formulations. The glycerol is central to all lipids known as triglycerides.
103 "Amino Acids", are molecules particularly important in biochemistry. Amino acids serve as the building blocks of proteins, which are linear chains of amino acids. Amino acids can be linked together in varying sequences to form a vast variety of proteins. Twenty amino acids are standard amino acids, and these 20 are encoded by the universal genetic code. Nine standard amino acids are called "essential" for humans because they cannot be created from other compounds by the human body, and so must be taken in as food. Amino acids are important in nutrition and are commonly used in nutrition supplements, fertilizers, food technology and industry.
104 Phytochemicals: chemical compounds that occur naturally in plants (phyto means "plant" in Greek), are responsible for color and organoleptic properties, such as the deep purple of blueberries and smell of garlic. The term is generally used to refer to those chemicals that may have biological significance but are not established as essential nutrients. Scientists estimate that there may be as many as 10,000 different phytochemicals having the potential to affect diseases such as cancer, stroke or metabolic syndrome. Although certain phytochemicals are available as dietary supplements, some scientists speculate that potential health benefits may best derive from consumption of whole foods.
105 Ibid.
106 "Science", *Webster's College Dictionary*, Merriam Webster, 1977
107 "Physical Fitness", *Wikipedia*, op. cit.
108 Ibid.
109 "Cortisol", *Wikipedia*, op. cit. Cortisol: is a steroid hormone produced by the adrenal gland. It is released in response to stress.
110 "Opioid peptides", *Wikipedia*, op. cit., are short sequences of amino acids that bind to opioid receptors in the brain; opiates and opioids mimic the effect of these peptides. Opioid peptides may be produced by the body itself, for example endorphins. The effects of these peptides vary, but they all resemble opiates. Brain opioid peptide systems are known to play an important role in motivation, emotion, attachment behavior, the response to stress and pain, and the control of food intake.
111 "Physical Fitness", *Wikipedia*, op. cit.
112 Ibid.
113 Ibid.
114 "Endorphins", *Wikipedia*, op. cit. Endorphins are opioid peptides that function as neurotransmitters. They are produced by the pituitary gland and the hypothalamus

in vertebrates during exercise, excitement, pain, consumption of spicy food, love and orgasm, and they resemble the opiates in their abilities to produce analgesia and a feeling of well-being. The term "endorphin rush" has been adopted in popular speech to refer to feelings of exhilaration brought on by pain, danger, or other forms of stress. When a nerve impulse reaches the spinal cord, endorphins that prevent nerve cells from releasing more pain signals are released. Immediately after injury, endorphins allow animals to feel a sense of power and control over themselves that allows them to persist with activity for an extended time.

[115] Ibid.
[116] Ibid.
[117] "Amenorrhea", *Wikpedia*, op. cit. Amenorrhoea is the absence of a menstrual period in a woman of reproductive age. Physiological states of amenorrhoea are seen during pregnancy and breastfeeding. Also, delay in pubertal development will lead to amenorrhoea.
[118] "Physical fitness", *Wikpedia*, op. cit.
[119] Bernie Keating, *Rational Market Economics: A Compass for the Beginning Investor*, Author House, Bloomington, IN, 2011
[120] *Websters Collegiate Dictionary*, Merriam Webster, 1977
[121] Adam Smith, *An Inquiry Into the Nature and Causes of the Wealth of Nations*, 1776. In his book, Smith explained that the fragments of social activity fitted together into a cohesive whole. The result was a blueprint for a new philosophy called economics. His theory led to a doctrine of "Laissez Faire." To him, the least government is best; all governments are spendthrift, irresponsible and unproductive. He was the ultimate proponent of the "Free Market."
[122] John Maynard Keynes, a Cambridge professor, was an economist during the period between the two World Wars. In 1935 he published *The General Theory of Employment, Interest, and Money*, which argues that the business cycle in the economy could stand still in a depression like a ship becalmed, and might need a nudge from government to get moving again. The pros and cons of the Keynesian theory are still being debated.
[123] "Economics", *Wikipedia*, op. cit.
[124] "Weather", *Encyclopedia Britannica*, 2007 Deluxe Edition
[125] Ibid.
[126] Ibid.
[127] Ibid.
[128] "Algorithms", *Wikipedia* 2007
[129] "Weather Forecasting", *Wikipedia*, op. cit.
[130] "Photovoltaic systems", *Wikipedia*, op. cit.
[131] "Conservation of Energy", *Wikipedia*, op. cit.
[132] Ibid.
[133] "Fossil Fuel", *Wikipedia*, op. cit.
[134] "Oil and Gas Journal", *Wikipedia*, op. cit.
[135] "Fossil Fuel", *Wikipedia*, op. cit.
[136] "Exxon's Big Bet on Shale Gas", *Fortune* Magazine, April 30, 2012, pg. 75
[137] "Hydraulic Fracturing", *Wikipedia*, op. cit.
[138] "Internal Combustion Engine", *Wikipedia*, op. cit.

139 Ibid.
140 Ibid.
141 "Solar Parks and Farms", *Wikipedia*, op. cit.
142 "Wind Power", *Wikipedia*, op. cit.
143 Ibid.
144 Ibid.
145 Ibid.
146 "Tidal Power", *Wikipedia*, op. cit.
147 Ibid.
148 "Malthus", *Wikipedia*, op. cit.
149 "Population", *Wikipedia*, op. cit.
150 "Peak Water", *Wikipedia*, op. cit.
151 Ibid.
152 Ibid.
153 Ibid.
154 Ibid.
155 Ibid.
156 Ibid.
157 Ibid.
158 Ibid.
159 "Water Desaltination", Wikipedia, op. cit.
160 Ibid.
161 Ibid.
162 Ibid.
163 "Water Desaltination", *Wikipedia*, op. cit.
164 "Discovery and Exploration of the Solar System", *Wikipedia*, op. cit.
165 "NASA" *Wikipedia*, op. cit.
166 "Discovery ...", *Wikipedia*, op. cit.
167 Ibid.
168 "Space Exploration", *Wikipedia*, op. cit.
169 "Sociology", *Wikipedia*, op. cit.
170 Adam Smith, *An Inquiry into the Nature and Causes of the Wealth of Nations*, 1776
171 Ibid.
172 Ibid.
173 Fareed Zakaria, *Time Magazine*, April 2, 2012, pg. 18
174 "Sociology', *Wikipedia*, op. cit.
175 "Einstein", *Wikipedia*, op. cit.
176 "Hans Albert Einstein", *Wikipedia*, op. cit.
177 "Einstein", *Wikipedia*, op. cit.
178 Ibid.
179 "Photoelectric Effect", *Wikipedia*, op. cit.
180 "Brownian Motion", *Wikipedia*, op. cit.
181 Einstein, Albert, *Fundamental Ideas and Problems of the Theory of Relativity*, Nobel Lectures, Physics 1901-1921, Amsterdam: Elsevier Publishing Company, retrieved 2007.
182 Ibid.

183 "Special Relativity", *Wikipedia*, op. cit.
184 "Mass-Energy Equivalence", *Wikipedia*, op. cit.
185 "Speed of Light", *Wikipedia*, op. cit.
186 Albert Einstein, *Relativity: The Special and General Theory*, first published in Great Britain by Methuen & Co. LTD, 1920, printed in the United States, Penguin, NY, 2006. Pg. viii. These observations were contained in the Introduction by Nigel Calder, Cambridge University. This book was written by Albert Einstein in 1916 when he was 37 years of age.
187 "Spacetime", *Wikipedia*, op. cit.
188 Albert Einstein, *Relativity: The Special and General Theory*, op. cit. pg. xxii
189 Ibid.
190 "Electromagnetism", *Wikipedia*, op. cit.
191 "Electromagnetic Radiation", *Wikipedia*, op. cit.
192 "James Clerk Maxwell", *Wikipedia*, op. cit.
193 "Radio Waves", *Wikipedia*, op. cit.
194 "Microwaves", *Wikipedia*, op. cit.
195 "Visible Spectrum", *Wikipedia*, op. cit.
196 "Ultraviolet", *Wikipedia*, op. cit.
197 "X-ray", *Wikipedia*, op. cit.
198 "Gamma Rays", *Wikipedia*, op. cit.
199 "Nuclear Force", *Wikipedia*, op. cit.
200 "Critical Mass", *Wikipedia*, op. cit.
201 Ibid.
202 "H Bomb", *Wikipedia*, op. cit.
203 "Nuclear Weak Force, *Wikipedia*, op. cit.
204 "Higgs ahoy!" *The Economist*, December 17th, 2011, pg. 18
205 "Introduction to Quantum Mechanics", *Wikipedia*, op. cit.
206 Alan Lightman, "The Accidental Universe", *Harper's Magazine*, December 2011, pg. 40
207 "String Theory", *Wikipedia*, op. cit.
208 Alan Lightman, op. cit. pg. 38
209 "Standard Model of Particle Physics", *Wikipedia*, op. cit.
210 "A Pallette of Particles", *American Scientist* Magazine, Sigma Xi, Vol. 100, March-April 2012, pg. 146
211 Ibid.
212 "Philosophy", Wikipedia, the free encyclopedia, 2011.
213 Philosophy", Collier's Encyclopedia, Vol. 15, P.F. Collier & Son Corp. 1960, pg 312
214 "Philosophy", *Wikipedia*, op. cit.
215 "Philosophy", Collier's, op. cit. pg. 313
216 "Philosophy, *Encyclopedia Britannica*, 2007 Deluxe Edition, Chicago
217 James Gorman, *The New York Times*, Carried in Modesto Bee 2/29/2012

www.ingramcontent.com/pod-product-compliance
Lightning Source LLC
Chambersburg PA
CBHW031049180526
45163CB00002BA/755